FOUNDATION MATHS

FOR GCSE AND STANDARD GRADE

ROBERT GLEN

HEINEMANN

L

Acknowledgements

The publisher would like to thank the following for
permission to reproduce photographs:

J Allan Cash (pp 4, 46, 135, 144 – West Indian
    girl, 177),
Allsport (pp 45, 84, 157, 180, 208),
Barnaby's Picture Library (pp 144 – white girl),
British Rail (p 158),
Alan Daykin (p 133),
Fiat UK Ltd (p217),
Sally and Richard Greenhill (pp 101, 108, 207),
Hulton Picture Library (p 215),
William J Howes (p 214),
Kobal Collection (p 131),
Sporting Pictures UK Ltd (pp 71 – all photos,
    p 87),
Topham (p 164),
Vauxhall Motors Ltd (p 114),
Mark Williams (p 160).

Heinemann Educational
a division of Heinemann Educational Books Ltd
Halley Court, Jordan Hill, Oxford OX2 8EJ

OXFORD LONDON EDINBURGH
MADRID ATHENS BOLOGNA PARIS
MELBOURNE SYDNEY AUCKLAND SINGAPORE TOKYO
IBADAN NAIROBI HARARE GABORONE
PORTSMOUTH NH (USA)

ISBN 0 435 50309 X

Designed by D. P. Press Ltd., Sevenoaks
Printed and bound in Great Britain by
Scotprint Ltd, Musselburgh

Preface

To the Student

Maths plays a much bigger part in your life than you might think. If you flick through this book, you'll find examples of Maths being used in sport and games, holidays and travel, at work and at home.

Think about going on holiday. If you go by coach or train, how do you work out the best journey? If you go by car, how do you plan your route? How do you estimate how long the car journey will take? If you go abroad, how do you get the best deal when exchanging your money? You may have solved problems like these without realising you were using Maths. Can you think of other problems you have solved which might involve Maths?

When you leave school, you will face new situations and challenges. This book will help you with some of those too – saving money, applying for jobs, understanding bills, and buying a car for example. Can you think of other situations in which you might need Maths?

This book covers all the skills you need for your exam. It also contains lively photos and illustrations to help make your studies more enjoyable. As you work through the examples and problems, I hope that you will learn to use Maths to solve problems of your own. Your Maths will then be useful long after you have taken your exam.

Good luck.

To the Teacher

Foundation Maths for GCSE and Standard Grade covers List 1 of the GCSE National Criteria for Maths, the syllabuses of the various examining groups at this level, and the Foundation Level requirements of Scottish Standard Grade.

I have written this book because as a teacher, I felt that no other textbook addressed a central and perennial problem; how to make Maths relevant, enjoyable, and useful to students at this level. While many existing texts include a selection of relevant applications, which follow on from the mathematical skills, I felt that none were based in the contexts of students' everyday experience, allowing the Maths to develop from familiar situations. As far as possible, I have tried to begin each unit of work with a familiar situation, so that the mathematical skills follow 'in context'.

I have presented the units in an order which enables students to work through the book sequentially, without becoming bored with a particular topic. For example, the longer topics (*Time, Money, Fractions*) are divided into shorter sections, and the numerical topics (*Whole numbers, Fractions, Decimals,* etc) are

separated with sections on *Shape, Angles,* and *Area* interspersed.

Each unit of work has a short descriptive heading which is a guideline to the mathematical content. This will help you to extract material on a particular topic, enabling you to cover the work in a different order if you wish.

The last section of the book, *Problems from the papers*, provides a wide range of mathematical topics, all based on newspaper articles and adverts. Many of these combine skills from different areas of Maths, so this section might be most profitably tackled at the end of the course.

The approach throughout is generally of an investigative nature, and many of the units can readily be extended, and practical work developed. Suggestions for further exercises and practical work are included in the Teacher's Notes and Answers. The symbol **T** has been used in this book where simple practical work is suggested and where students are specifically asked to check their working with you.

Robert Glen

Contents

Numbers all around

Study this street scene carefully.

There are lots of numbers in this picture.

Examples

The 7 is the plumber's address.
The 28 tells us the size of the TV screen
in inches.

1 Find each of these numbers in the picture.
   Explain their meanings as in the examples
   above.
   (a) 90   (b) 400
   (c) 604 1523   (d) $5\frac{1}{2}$
   (e) The 24 in the plumber's window.

2 The woman in the picture is just about to put in
   some films for developing.
   When should she return to get her prints?

3 For how many years has the plumber's firm
   been going?

4 Mr Laird's car has an engine size 400 ml bigger
   than his van.
   What engine size is his car?

5 How much would you expect to pay for the
   28 inch TV?

6 Draw rectangles in your notebook to show the
   four different sizes of print available.
   Arrange these in order, according to how much
   paper they use up.

On the button

Many of us use keyboards with numbers on them every day.

1 Match each keyboard to the person who is using it.

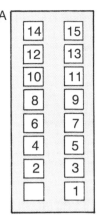

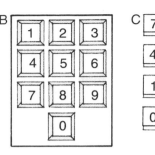

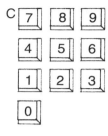

Example
Picture A shows the keyboard inside a lift. Picture F shows a small boy using a lift.

We often read numbers on a calculator display.

Example
This calculator display reads two hundred and eighty-five thousand.

285000

---

2 Write down in words the reading on each of these calculator displays.

(a) **348**

(b) **1795**

(c) **24186**

(d) **39020**

(e) **752000**

(f) **1475000**

T 3 Your teacher will give you a sheet of blank calculator displays.
Draw each of these numbers as the calculator would display it.
(a) three hundred and twenty-eight
(b) five thousand, one hundred and seventy-nine
(c) twenty-five thousand
(d) five million
(e) three million, eight hundred thousand

4 Show how the calculator would display answers to these calculations.
Check by using a calculator.
(a) $999 + 1$ (b) $10\ 000 - 1$
(c) 5.4×10 (d) 5.4×100
(e) $736 \div 10$ (f) $736 \div 100$

Scales, clocks and dials

1 Here are some people using different scales
(number lines).
Match up each person to the scale they are
reading.
Write down the reading on each scale as
accurately as you can.

> **Example**
> Picture A shows someone reading the
> temperature. Scale H shows a
> thermometer. The thermometer reads
> about 30°C (about 85°F).

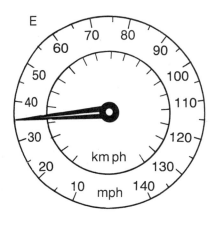

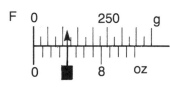

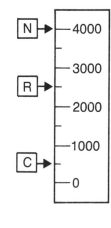

2 Copy this table.

S							
24	2500	180	2600	2800	360	400	13

(a) Find each number on one of the five scales below.
Above each number write the letter that appears on the scale.

(b) Write out the letters in reverse order to find the name of a
well-known film.

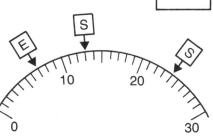

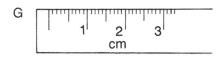

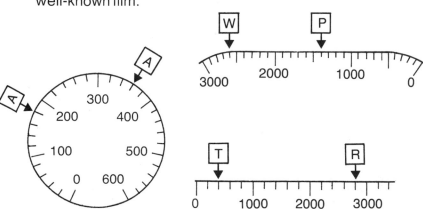

3

3 Study these 8 number lines carefully.

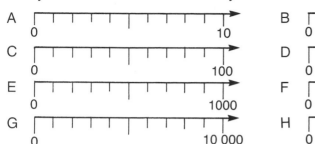

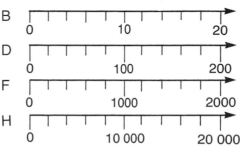

Copy each of the number lines and extend it across the full width
of your notebook.

Example

Line A

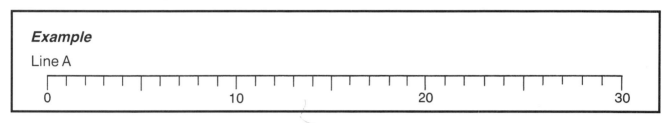

---

4 Mark each of these numbers on one of the number lines you have drawn.
Choose the number line which shows the number most accurately.

Example

The number of weeks in a year.
This number is 52. Line A would not be long enough.
Line B would show 52 most accurately.

Number of weeks
in a year

Line B

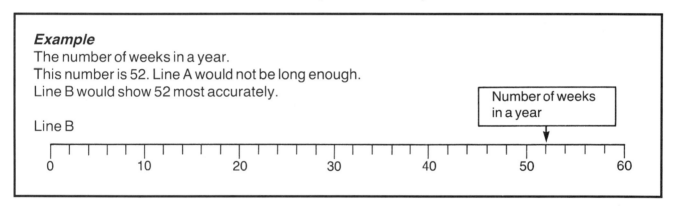

(a) Your age.
(b) Your mother or father's age.
(c) Your grandmother or grandfather's age.
(d) The height of Mount Everest measured in
   metres.

(e) The cost of a middle-of-the-range new car.
   (Choose one from the newspaper.)
(f) Today's temperature in °C.
(g) Today's temperature in °F.
(h) The annual salary in pounds of an
   average wage earner.
(i) The population of your town or village.
(j) The number of seconds in 12 hours.
(k) The number of squares
   like this which would
   cover this page.

(l) The number of students in your class.
(m) The number of students in your school.

Making the headlines

Tony works for a local newspaper.

1 Rewrite each of these headlines with 'nice round numbers'.

(a) **Vandals cause £7950 fire**

(b) **Town unemployed rises to 25 147**

(c) **Local man fined £497**

(d) **35 245 see United win again**

(e) **Villager wins £485 476**

(f) **Council to build 39 new houses**

(g) **Martha's 387 mile charity hike**

(h) **987 500 attend pop festival**

2 (a) Mr Anwar is paid £14 895 per annum.
     How much is this, rounded to the nearest
     thousand pounds?
   (b) His salary is paid in equal monthly
     instalments.
     How much is he paid each month, rounded
     to the nearest £100?

3 (a) How much did the four cleaners win
     altogether, rounded to the nearest
     £100 000?

   (b) Exactly how much was each person's
     share?
   (c) Complete this headline about one of the
     cleaners, using a rounded number.

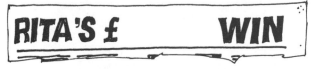

RITA'S £   WIN

4 72 436 fans attended Wham's farewell
   concert at Wembley in 1986.
   How many is this, rounded to the nearest
   thousand?

Rough and ready

Example

I wonder if £10 is enough for these?

£2.58
52p
£1.89
47p
£3.96

It comes to about £9.50. £10 should cover it.

£2.50
50p
£2
50p
£4

1   Round each item on these bills to the nearest 50p as in the
    example above.
    Work out a rough total for each bill.
    Then work out an accurate total using a calculator.

(a)	£0.94	(b)	£4.75	(c)	£0.59	(d)	£2.75	(e)	£8.42	(f)	£10.75
	£2.08		£1.96		£0.93		£1.87		£5.99		£ 4.99
	£1.47		£3.24		£0.12		£3.14		£1.46		£ 3.25
	£0.39		£2.83		£0.36				£4.84		£ 6.85
					£0.87				£3.15		£ 7.25

Here is how to estimate answers to multiplying and dividing problems:

Example

I managed to save £3.80 last week. How long will it take me to save enough for this personal stereo?

£39

Number of weeks is   £39 ÷ £3.80
Roughly   £40 ÷ £4
                              = 10 weeks

Check 10 × £3.80 to see how close to
£39 he would be in 10 weeks,
                              = £38.

2   Do a rough estimate first in each of these
    problems, then work out the accurate answer.
    (a) Ahmed buys twelve 18p stamps.
        How much will they cost altogether?
    (b) 4 girls share a win of £7960.
        How much does each girl get?
    (c) What is the cost of 17 cut-price records at
        £2.15 each?
    (d) A brick is 23 cm long.
        How far will 175 bricks stretch?

    (e) The netball team wants to buy 7 strips at
        £18.95 each.
        How much will they have to raise?
    (f) A school party of 29 children is charged
        £27.55 for entrance to the zoo.
        How much is this for each child?

What comes next?

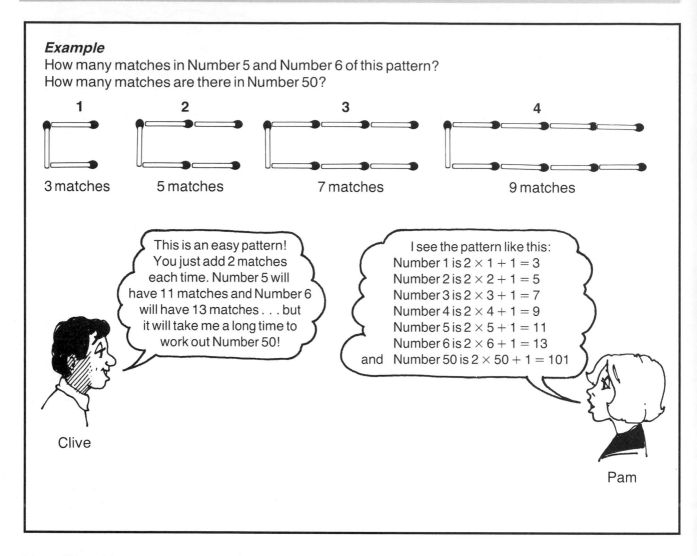

Example
How many matches in Number 5 and Number 6 of this pattern?
How many matches are there in Number 50?

1   **2**   **3**   **4**

3 matches   5 matches   7 matches   9 matches

This is an easy pattern!
You just add 2 matches
each time. Number 5 will
have 11 matches and Number 6
will have 13 matches . . . but
it will take me a long time to
work out Number 50!

Clive

I see the pattern like this:
Number 1 is $2 \times 1 + 1 = 3$
Number 2 is $2 \times 2 + 1 = 5$
Number 3 is $2 \times 3 + 1 = 7$
Number 4 is $2 \times 4 + 1 = 9$
Number 5 is $2 \times 5 + 1 = 11$
Number 6 is $2 \times 6 + 1 = 13$
and   Number 50 is $2 \times 50 + 1 = 101$

Pam

You will be able to answer some of these questions like Clive.
For some questions you must think out the pattern like Pam.

1 (a) What are the addresses of the 5th and 6th
houses on the street?

Willow Street

(b) What is the address of the 15th house on
the street?
(c) The addresses on the other side of the
street are odd numbers starting from 1.
What are the addresses of the first 5
houses on the other side?
(d) What is the address of the 15th house on
the other side?

2 (a) Copy these 3 pictures of stacks of tins.

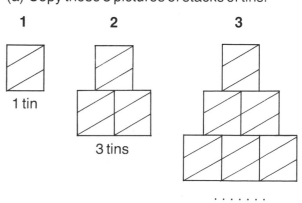

1   **2**   **3**

1 tin

3 tins

. . . .

(b) Draw the next 2 stacks in the pattern.
(c) How many tins are in stack number 6 and
stack number 7?

3 (a) Draw the next 2 squares in this pattern.

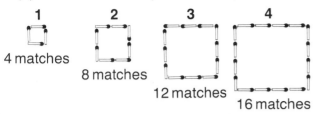

1
4 matches

2
8 matches

3
12 matches

4
16 matches

(b) How many matches are there in square number 7 and square number 8?

(c) How many matches are in square number 20?

4 (a) Draw the next 2 squares in this pattern of slabs.

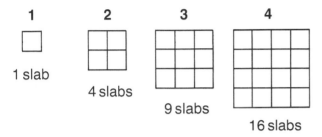

1
1 slab

2
4 slabs

3
9 slabs

4
16 slabs

(b) How many slabs are there in square 7 and square 8?

(c) How many slabs are there in square 20?

5 This pattern is made by tables along the wall of a cafe.

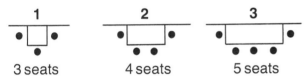

1
3 seats

2
4 seats

3
5 seats

(a) How many seats are there at table 4 and table 5?

(b) How many seats are there at table 10?

6 (a) Draw the next 2 flowerbeds in this pattern.

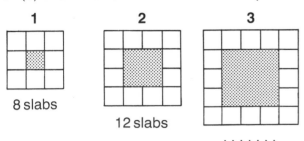

1
8 slabs

2
12 slabs

3

(b) How many slabs are there in gardens 6 and 7?

(c) How many slabs are there in garden 10?

7 (a) Draw the next 2 frameworks in this pattern of triangles.

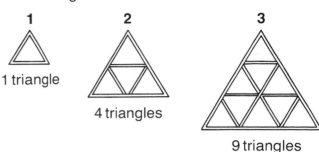

1
1 triangle

2
4 triangles

3
9 triangles

(b) How many triangles are there in number 6 and number 7?

(c) How many triangles are there in number 10?

8 (a) Draw number 4 and number 5 in this pattern of slabs.

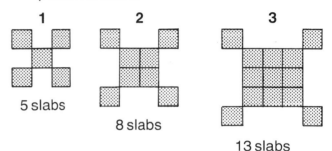

1
5 slabs

2
8 slabs

3
13 slabs

(b) How many slabs are there in number 6 and number 7?

(c) How many slabs are there in number 15?

9 Put the next 3 numbers in each of these patterns.
(a) 2, 4, 8, 16 . . .
(b) 3, 6, 9, 12 . . .
(c) 1, 2, 6, 24 . . .
(d) 0, 3, 8, 15, 24 . . .
(e) 1, 3, 6, 10 . . .
(f) 1, 8, 27 . . .
(g) 1, 4, 9, 16 . . .
(h) 7, 14, 21, 28 . . .
(i) 1, 2, 4, 7 . . .
(j) 0, 2, 6, 12 . . .
(k) 1, 4, 16, 64 . . .

In your head

Examples

My baby needs one for each day of the week.

That will be £7 less 7p. That's £6.93.

I've scored treble 17. That's three 20s less three 3s.

That's 60 − 9. Which is 51.

Work out answers to these problems in your head like the people in the pictures above.

1 A jar of coffee costs £2.99. How much do 3 jars cost?

2 What change do you get from £1 when you buy 5 of these stamps?

3 Mrs Binns buys 2 pints of milk per day, 7 days a week. The dairy charges 25p per pint. How much is Mrs Binns' weekly bill?

4 What is the total score from these 3 darts?

5 Two 175 ml bottles of medicine are filled from a large bottle containing 750 ml.
   (a) How much medicine is left in the large bottle?
   (b) How many more small bottles could be filled from what is left in the large bottle?

6 How much would you pay for 1 lb of sausages and one dozen eggs?

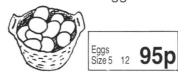

Eggs Size 5 12 **95p**

Pork and Beef Sausages (loose) per lb **89p**

7 A van driver finds that the journey from Milltown to Greenfield to Springburn is exactly the same distance as Milltown to Hillfoot to Springburn.
What is the distance between Hillfoot and Springburn?

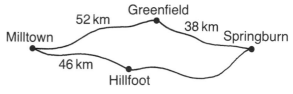

8 (a) The dealer still makes a profit of £35 on each TV set.
   How much did each set cost him?
   (b) A hotel proprietor buys 7 of the sets.
   How much did they cost him?

£119

£20 off MARKED PRICES

Using your calculator

Use a calculator for all these questions.
Write down the calculation you are doing for each question and write a sentence to explain your answer.

1 This table shows the daily takings at a supermarket for one week.

Monday	Tuesday	Wednesday
£9544	£11 746	£10 847
Thursday	**Friday**	**Saturday**
£25 640	£28 715	£37 415

(a) Calculate the total takings for the week.
(b) The target for the week is £125 000.
   How far short of the target did the takings fall that week?

2 Here is the mileage reading on a car before a journey.

| 2 7 7 5 6 |

Here is the reading after the journey.

| 2 7 9 9 1 |

(a) How far did the car travel?
(b) The journey took 5 hours.
   What was the car's average speed in mph?

3

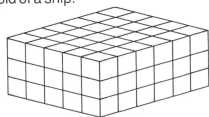

New computers sell at the rate of one every minute for the first week.

How many new computers are sold in the first week?

4 The diagram shows crates stacked together in the hold of a ship.

(a) How many crates are there in the top layer?
(b) How many layers are there?
(c) How many crates are there altogether?

5 A firm made a loss of £1 750 000 in 1985 and a profit of £2½ million in 1986.
How much of an improvement is this?

6 Study this graph and use it to answer the questions.

First year attendance Friday

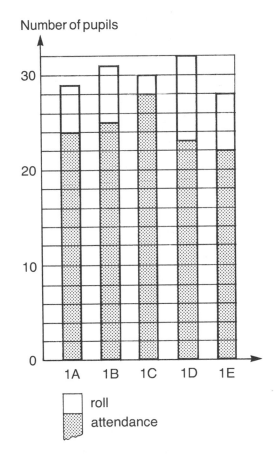

(a) How many first year pupils were present last Friday?
(b) How many were absent?
(c) What is the total roll of the first year?
(d) What is the average roll per class?
(e) What was the average attendance per class last Friday?

10

A busy sales rep

Pauline works for a publishing company.
She visits schools showing teachers which
books are available.
This map shows the main towns and cities in
Pauline's area.

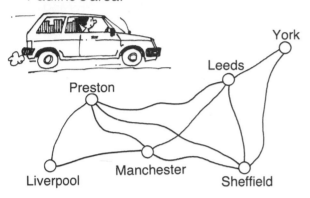

The distance chart below shows the mileage
between all the towns in her area.

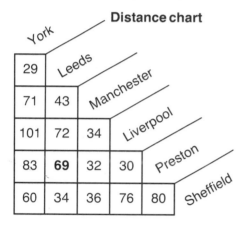

Example
The distance between Preston and Leeds
is 69 miles.

1 Write down the distance between each pair of
towns.
  (a) Leeds and Liverpool
  (b) York and Leeds
  (c) Preston and Sheffield
  (d) Manchester and York
  (e) Preston and Liverpool
  (f) Manchester and Liverpool
Check with your teacher that your answers
are correct before you go on.

2 Pauline lives in Preston. On Monday she
visited schools in Manchester and Liverpool.
Copy and complete her mileage table for
Monday.
(Use your answers to 1(e) and (f)).

Monday

Preston – Manchester	Manchester – Liverpool
<u>32</u> miles	___ miles
Liverpool – Preston	**Total mileage**
___ miles	___ miles

3 On Tuesday Pauline followed this route:
Preston ➤ Manchester ➤ Leeds ➤ Preston
  (a) Make out her mileage table for Tuesday.
  (b) This is the mileage reading on Pauline's
  car before she left home on Tuesday
  morning.

  | 4 6 7 1 8 |

  What was the mileage reading when she
  arrived back home?

4 On Wednesday Pauline travelled from Preston
to one of the main towns in her area and then
on to Sheffield where she stayed overnight.

Mileage before leaving
Preston | 4 6 8 6 5 |

Mileage on arrival at
Sheffield | 4 7 0 0 8 |

What is her total mileage for Wednesday?
Copy her mileage table for Wednesday and fill
in all the blanks.

Wednesday

Preston –	miles
. . . . – Sheffield	miles
Total mileage	miles

Scoring a century

In this game you throw a die on to the hexagons on the board.

The rules for playing the game are given in the flowchart below.

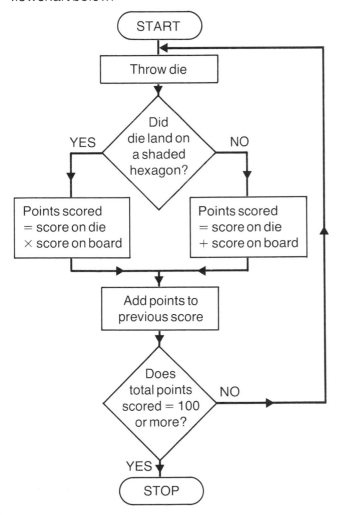

START

Throw die

Did die land on a shaded hexagon?

YES → Points scored = score on die × score on board

NO → Points scored = score on die + score on board

Add points to previous score

Does total points scored = 100 or more? → NO

YES → STOP

Example

The die lands on **5** and shows ⚃

5 is shaded so multiply.

Score is 4 × 5, which is 20.

1 Copy and complete this table to show all the possible points you can score in this game.

Score on board

Score on die	④	**5**	⑥	**7**	⑧	**9**	⑩
•							
••							
•••							
••••	8	20	10	28	12	36	14
•••••							
••••••							

2 What is the maximum possible score with one throw?
What is the minimum?

3 Copy this table, then fill in the points for each person, using your table from question 1.

Reeta

Throw	Die	Board	Points
1	⚄	⑩	16
2	⚂	**7**	
3	⚃	**9**	
4	⚁	⑥	
		Total points	90

Clive

Throw	Die	Board	Points
1	⚄	**7**	
2	⚂	⑧	
3	⚅	**9**	
4	⚄	**5**	
		Total points	

Who finished the game?
Who got closest to 100 in 4 throws?

Getting into shape

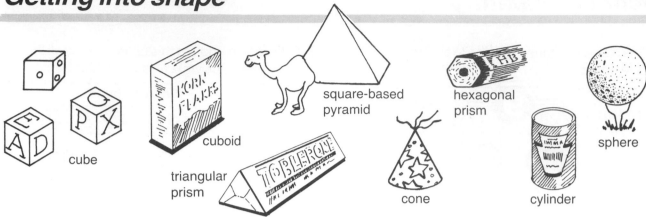

square-based pyramid

hexagonal prism

cube

cuboid

triangular prism

sphere

cone

cylinder

1 Give one real-life example for at least 6 of the shapes in the picture above. (Do not use the examples in the picture.)

2 Write down the name of each numbered shape in this model of a factory.

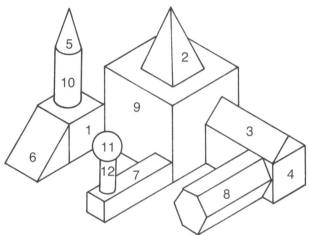

There are good reasons for most everyday objects being the shape they are.

Example   Rolling pin

Usual shape: cylinder

Rolls smoothly. No edges to dig into dough.

New shape: cuboid

Would not roll. Edges would dig into dough and cut it up.

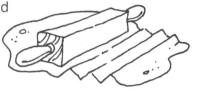

---

3 Here are some suggestions for new shapes.
Write down the drawbacks of the new shapes and say why the usual shapes are better.

(a) Football

Cube

(b) Tin of beans

Triangular prism

(c) Die

Sphere

(d) Car wheel

Hexagonal prism

Making shapes

Karen helps in a nursery school as part of her community project.
One of her jobs is to help the children make up shapes.

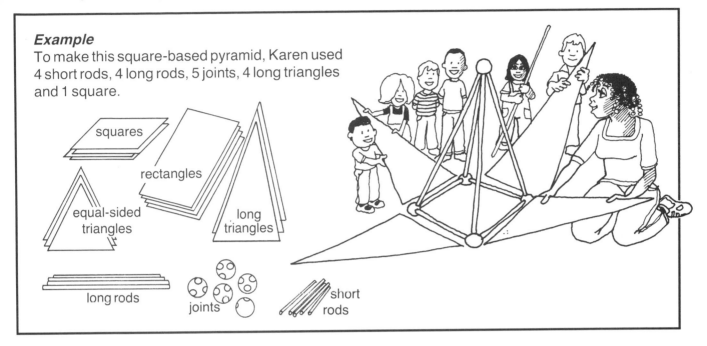

Example
To make this square-based pyramid, Karen used
4 short rods, 4 long rods, 5 joints, 4 long triangles
and 1 square.

squares

rectangles

equal-sided triangles

long triangles

long rods

joints

short rods

1 What would Karen need to make each of the shapes below?
(Only the 'skeleton' is shown for each shape.)
Write your answers in a table like this:

Shape	Edges		Vertices	Faces			
	Short rods	Long rods	Joints	Squares	Rectangles	Long triangles	Equal-sided triangles
Cube	12	0	8	6			

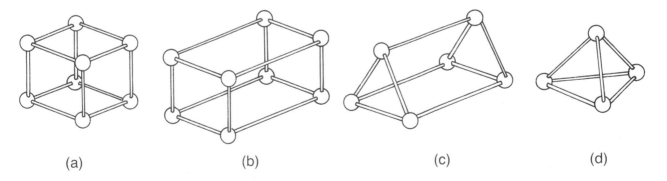

(a)　　　　　　(b)　　　　　　(c)　　　　　　(d)

2 Karen made some hats for the children, who were having a fancy
dress party.
What shapes did Karen cut out of cardboard to make these two
hats?

Make a small version of each hat yourself.

Tins and boxes

A 450 g tin of beans is 115 mm in height and 75 mm in diameter.
The tins are delivered to the shops in cartons like the one in the picture.

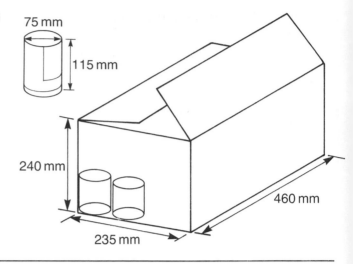

Example
How many tins will fit across the width of the carton?

Try 4: 4 × 75 mm = 300 mm. Too many.

Try 3: 3 × 75 mm = 225 mm. O.K.

Three tins will fit across the width.

---

1 (a) How many tins will fit along the length of the carton?
  (b) How many tins will cover the bottom of the carton? (The example above will help you with this.)
  (c) How many layers of tins will the carton hold?
  (d) How many tins will the carton hold?

Jason, would you stack the tins from the carton in a triangular display like this, please.

  (e) Draw the biggest triangle Jason could make from the tins in one carton.
  (f) What weight of beans are in this triangle?

2 One afternoon, during a quiet time, Jason thought he would try to find as many ways of stacking the tins from one carton as he could.

Example
Here is one way Jason found.

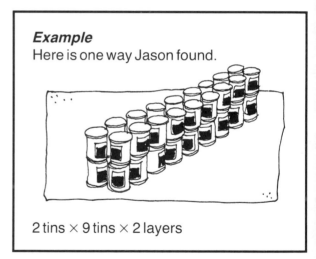

2 tins × 9 tins × 2 layers

Explain all the ways you can find.
The sketches below will give you some clues.

Making boxes

Maxine works in a factory which makes cardboard boxes.
She designs the net of the box, then the factory mass produces the boxes. (The net is a flat shape which folds up to make a box.)

Here are some of the jobs she was given last week.

1

Maxine, Minim Records want a box to hold a double cassette album, in 2 different styles.

I'll do one like this.
A

. . . and one like this.
B

A single cassette is 110 mm by 70 mm by 16 mm. I'll use Net A for the first style. This will be open at one end. Net B is for the second style.

(a) Copy the net of box A.
T (b) Fold and glue the net to make the open box. Check that 2 cassettes will fit into the box.
(c) Repeat for box B.

Net A

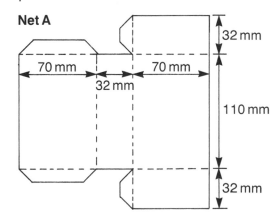

70 mm · 70 mm
32 mm
32 mm
110 mm
32 mm

Net B

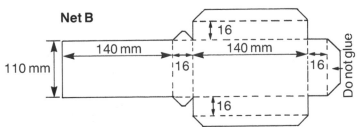

110 mm · 140 mm · 140 mm
16 · 16 · 16 · 16
Do not glue

2

NEW 'SUDS' CARTON

25 cm
10 cm · 15 cm

Maxine, design a carton to hold 24 of these soap powder boxes. The boxes can stand any way in the carton.

(a) Find one way of stacking 24 boxes.
(b) Work out the length, width and height of the carton you would need to hold 24 boxes.
(c) Draw the net of this carton using a scale of 1:10. (1 cm on your drawing stands for 10 cm on the full-size box.)
(d) Fold and glue the net to make a model of the carton.
(e) Draw lines on the outside of your model to show how the 24 boxes will fit inside the carton.
(f) Maxine discovers that the factory already makes these boxes.

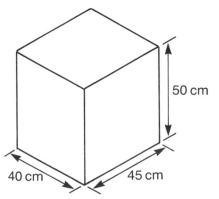

50 cm
40 cm · 45 cm

Explain how Maxine can use this to hold 24 boxes of soap powder.

Cube capers

1 Zahina put this ice cube into a measuring jar.

When the ice melted the water came up to the first mark.

The water from this cube came up higher.

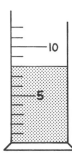

I know how high the water from each of these cubes will come.

(a) Write down the water levels for the last 2 cubes.
(b) Draw an ice cube which would make the water level come up to 125.

2 Count the cubes in each of these shapes.

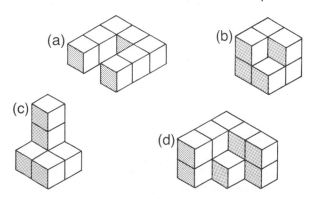

(a)

(b)

(c)

(d)

3 Sugar cubes are 15 mm long. The cubes are packed into a box like this:

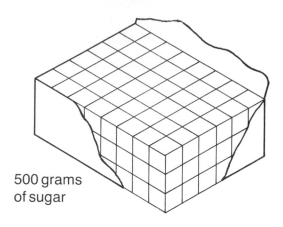

500 grams of sugar

(a) Work out the length, breadth and height of the box.
(b) How many cubes are packed into the box?
(c) Find at least one other way to pack the same number of cubes into a box.

Make a sketch of your new box to show how the cubes would be packed. (You will **T** need a sheet of isometric dot paper.)

Mark drinks 5 cups of tea every day.
He puts 3 sugar cubes in every cup.

(d) Mark's dad estimates that it will take Mark about 19 days to swallow 1 kilogram of sugar in his tea.
How close to the truth is this?

(e) Mark finds out that too much sugar is bad for you.
He decides to take only one sugar lump in each cup.
Roughly how many days does it now take Mark to swallow 1 kilogram of sugar?
How many kilograms of sugar does he now take in one year?

Points of view

How good are you at seeing things in 3 dimensions?

1 Which 3 of these pieces fit together to make the large cube?

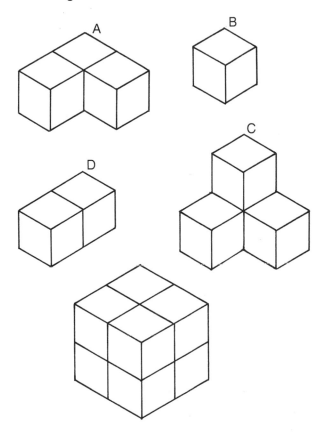

2 Who sees which view?

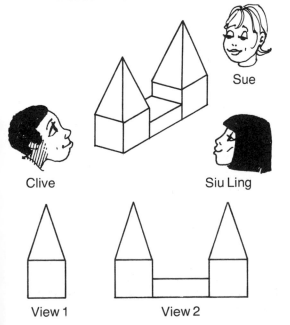

Sue

Clive

Siu Ling

View 1   View 2

3 Who sees which view?

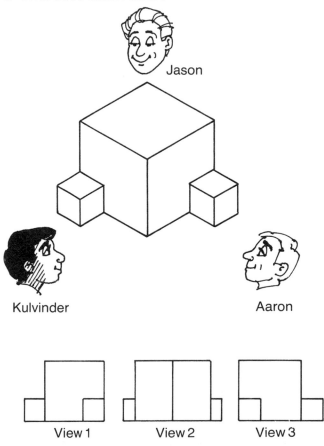

Jason

Kulvinder   Aaron

View 1   View 2   View 3

4 Pair off the pieces which could make the T-joint.
Which 2 pieces are the odd ones out?

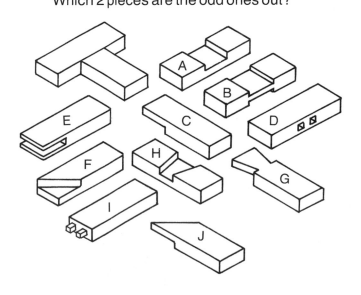

Mystery prize

In this TV quiz game the contestants do not know what is in each box – but you do.

> **Example**
> Jill picked box (E3) and won a set of wine glasses.

1 What did each of these contestants win?
  (a) Eleanor (C3) (b) Ahmed (D1)
  (c) Yvonne (C5) (d) Bob (D3)
  (e) Jason (E2) (f) Claire (A1)

2 Which box did each of these prizewinners choose?
  (a) John .... cake
  (b) Parminder .... cutlery
  (c) Ivan .... potato
  (d) Sibohann ...... lawnmower
  (e) Harry ... washing machine
  (f) Megan .... clothes peg

Pick a box

---

3 In the grid below letters of the alphabet are placed where the lines cross.

> **Examples**
> Position (8,2) takes us to letter J
> Position (2,8) takes us to letter C

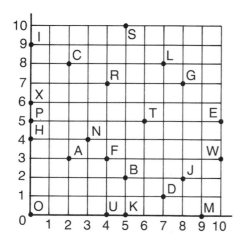

(a) Copy each of the tables below.

(3,4)	(0,0)	(0,0)	(3,4)	(7,1)	(7,8)
N	O			D	

(3,4)	(4,0)	(10,5)	(7,1)	(10,5)	(7,1)

(5,10)	(6,5)	(2,3)	(5,2)	(7,8)	(10,5)	(4,3)

(10,3)	(0,4)	(10,5)	(4,7)	(9,0)	(2,3)	(0,6)

(5,2)	(4,7)	(0,0)	(6,5)	(3,4)	(0,9)	(8,7)	(0,4)

(b) Fill in the correct letter for each position.
(c) Unjumble the letters in each table to find the names of 5 British towns.

Decode the message

This coding grid uses circles
and lines at different angles.

Examples
(4,0°) is the code for
letter N.
(3,60°) is the code for
letter E.

An agent knows that the code
sequence is (4,0°), (3,30°),
(2,60°), (1,90°), (1,120°),
(2,150°), (3,180°), (4,210°),
(4,240°), (3,270°), etc. twice
round the circle.

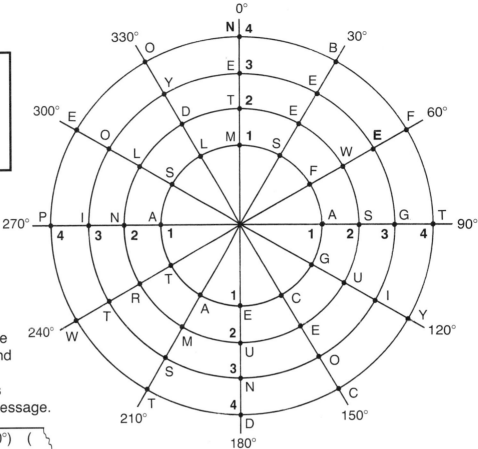

1 (a) Write out the whole code
sequence for twice round
the circle.

(b) Copy and complete this
table and decode the message.

(4,0°)	(3,30°)	(2,60°)	(
N	**E**		

2 Draw a grid like the one in question 3 on
page 19, using ½ cm squared paper.
Number the grid up to 20 in each direction.

Plot these points on the grid.
Use a ruler to join up the points in order as you
go along.

(a) (2,2) → (10,0) → (18,2) → (20,10) → (18,18) → (10,20) → (2,18) → (0,10) → (2,2)

(b) (6,1) → (14,19) → (14,14) → (19,14) → (1,6) → (6,6) → (6,1)

(c) (6,6) → (14,14)

(d) (14,1) → (6,19) → (6,14) → (1,14) → (19,6) → (14,6) → (14,1)

(e) (6,14) → (14,6)

(f) (9,4) → (11,16) → (10,17) → (9,16) → (11,4) → (10,3) → (9,4)

(g) (4,9) → (16,11) → (17,10) → (16,9) → (4,11) → (3,10) → (4,9)

Squares

Example

(a) Plot the points P(2,1), Q(5,1) and R(5,4)

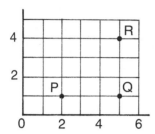

(b) Join up P → Q → R.

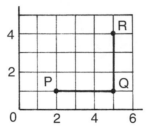

(c) PQRS is a square. Find the position of point S.

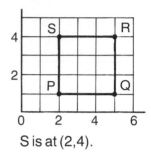

S is at (2,4).

Answer questions 1, 2 and 3 as in the example above.

1 A(2,6) B(2,2) C(6,2)
ABCD is a square.
Find the position of D.

2 K(1,2) L(1,7) M(6,7)
KLMN is a square.
Find the position of N.

3 E(2,2) F(4,2) G(4,4)
EFGH is a square.
Find the position of H.

Example

(a) Plot the points A(3,0), B(5,2) and C(3,4)

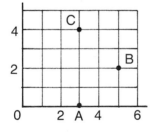

(b) Join up A → B → C.

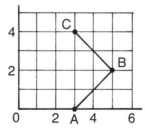

(c) ABCD is a square. Find the position of point D.

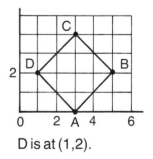

D is at (1,2).

Answer questions 4 to 6 as in the example above.

4 P(4,1) Q(7,4) R(4,7)
PQRS is a square.
Find the position of S.

5 K(4,0) L(0,4) M(4,8)
KLMN is a square.
Find the position of N.

6 E(8,4) F(5,7) G(2,4)
EFGH is a square.
Find the position of H.

For questions 7 and 8, use a grid drawn on 1 cm squared paper.

7 A is the point (1,1) and B is the point (4,1).
AB is one side of a square.
Work out the perimeter of the square.

8 L is the point (1,1) and N is the point (5,5).
The line LN is one diagonal of a square.
Work out the perimeter of the square.

Growing up

Example
Enlarge this diagram to 2 times its present size
and to 3 times its present size.

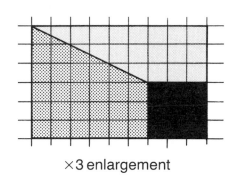

original size   ×2 enlargement   ×3 enlargement

1 Copy each of these diagrams.
Draw a ×2 enlargement and a ×3 enlargement of each.

(a) (b) (c) 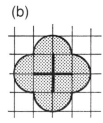 (d) 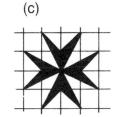 (e)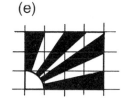

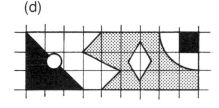

Example
This picture of a car is drawn on $\frac{1}{4}$ cm squared
paper.
William copied the picture one square at a
time on to $\frac{1}{2}$ cm squared paper.
The drawing on the right is a ×2 enlargment
of the small picture.

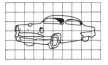
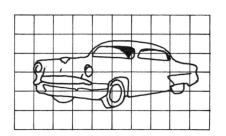

2 Draw a ×2 enlargement of the dog and the gymnast.

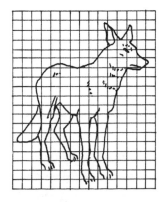
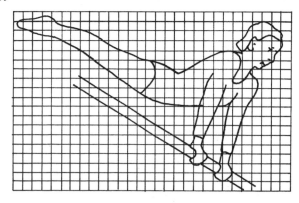

22

2 Here are the results of a survey of shoe sizes among fifth year pupils.
(a) Copy each table and fill in the 'Frequency' column.

Fifth year girls

Shoe size	Tally	Frequency
1	ⵏⵏ ⵏⵏ	10
2	ⵏⵏ ⵏⵏ I	11
3	ⵏⵏ ⵏⵏ ⵏⵏ IIII	
4	ⵏⵏ ⵏⵏ ⵏⵏ ⵏⵏ IIII	
5	ⵏⵏ ⵏⵏ ⵏⵏ	
6	ⵏⵏ ⵏⵏ IIII	
7	ⵏⵏ I	
8	I	

Fifth year boys

Shoe size	Tally	Frequency
4	IIII	
5	ⵏⵏ ⵏⵏ	
6	ⵏⵏ ⵏⵏ ⵏⵏ ⵏⵏ	
7	ⵏⵏ ⵏⵏ ⵏⵏ IIII	
8	ⵏⵏ ⵏⵏ ⵏⵏ ⵏⵏ ⵏⵏ I	
9	ⵏⵏ ⵏⵏ ⵏⵏ	
10	IIII	
11	II	

(b) Draw a bar graph to show the girls' shoe sizes.

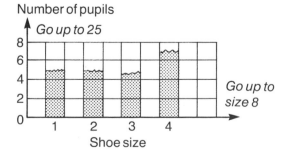

Number of pupils

Go up to 25

Go up to size 8

Shoe size

(c) Draw another bar graph to show the boys' shoe sizes.

(d) Next week 20 boys and 20 girls from the fifth year will go to the roller disco. The names will be picked from a hat. You have to phone the roller disco today to give a rough idea of how many of each size of roller boot they will want to hire. How many of each size would you order? (No-one has boots of their own.)

(e) Draw a bar graph of your order for girls' boots underneath your first graph. Repeat for your order for boys' boots.
(f) As a group project, make a survey of shoe sizes for all the pupils in your year.
(g) Answer questions (b), (c), (d) and (e) for pupils in your year.
T Your teacher will help you with the awkward calculations you will have to do in part (d).

Reflections

Example
Place a mirror along the
broken line and look
from this side.
You should see exactly the
same in the mirror as you
do when you take the mirror
away again.
The broken line is called a
line of symmetry of the
cat's face.

1 Copy or trace each of these shapes.
 Mark in all the lines of symmetry on each
 shape, using a mirror to help you.
 (Some shapes have no line of symmetry,
 some have one, some have two or more.)

2 Find 2 different ways to rearrange the
 ornaments on Grandma's mantelpiece so that
 the broken line is a line of symmetry.

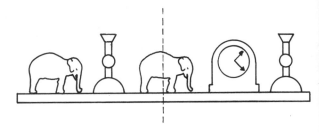

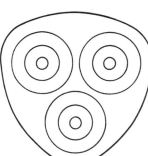

(a)

(b)

3 Copy each of the shapes below, then complete
 them so that the broken line is a line of
 symmetry.

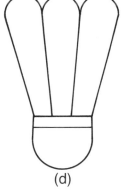

(c)

(d)

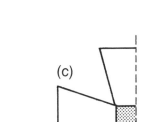

(b)

(c)

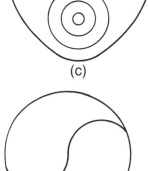

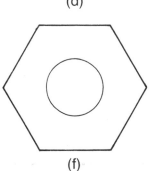

(e)

(f)

(a)

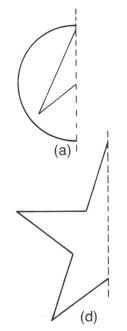

(d)

4 Each of these shapes has a lettered part and a numbered part.
Pair off the lettered parts and numbered parts to make shapes
with a line of symmetry. Draw every shape you find on $\frac{1}{2}$ cm
squared paper.

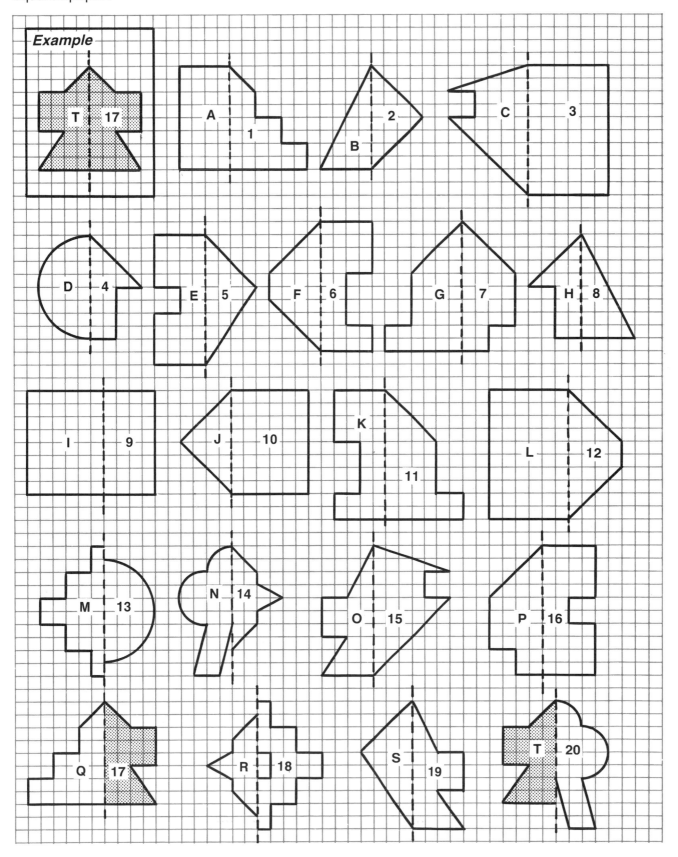

Angles everywhere

Example 1

The ramp is at an angle of 10° to the ground.

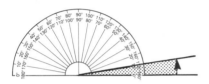

This is called an **acute angle**. (Less than a right angle).

Example 2

The dancer's body is at an angle of 90° to her legs.

This is called a **right angle**.

Example 3

The lid of the ring box has been turned through an angle of 120°.

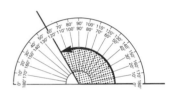

This is called an **obtuse angle**. (More than a right angle).

1 Measure the angle in each of the pictures below.
Say whether each is an acute angle, an obtuse angle or a right angle.

(a)

(b)

(c)

(d)

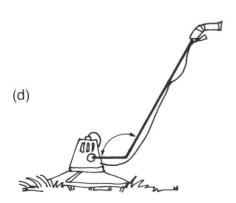

(e)

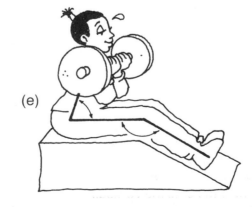

(f)

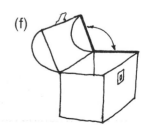

2 Without measuring, write down whether each shaded angle is
acute, right or obtuse, and estimate the size of the angle.
Then check your answer by measuring.

3 (a) Draw a clock with the hands at an angle of 60°.
   (b) Draw a sunbed set at an angle of 150°.
   (c) Draw a reading lamp set at 110° and 60°.
   (d) Draw a fishing rod held at 70° to the vertical.
   (e) Draw an aeroplane diving at an angle of 25° to the horizontal.

Zig-zag

Example

Angle A measures 25°.

Angle B measures 125°.

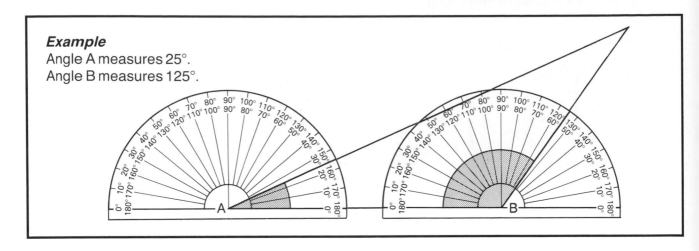

1 Measure the 3 angles of this triangle.

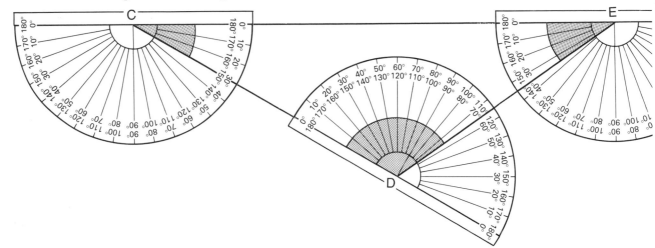

2 Now measure all the angles in the zig-zag using a protractor.

Snookered

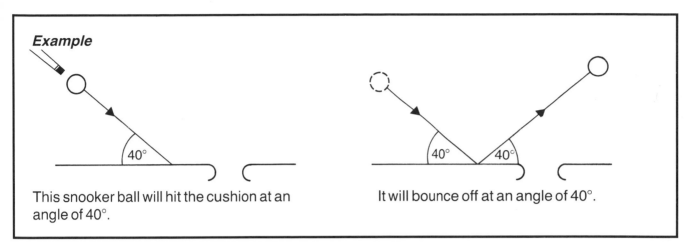

Example

This snooker ball will hit the cushion at an angle of 40°.

It will bounce off at an angle of 40°.

Copy each diagram. Draw a line to show where the ball will go after it bounces off the cushion. (All the angles are multiples of 5°.)

1   **2**   **3**   **4**

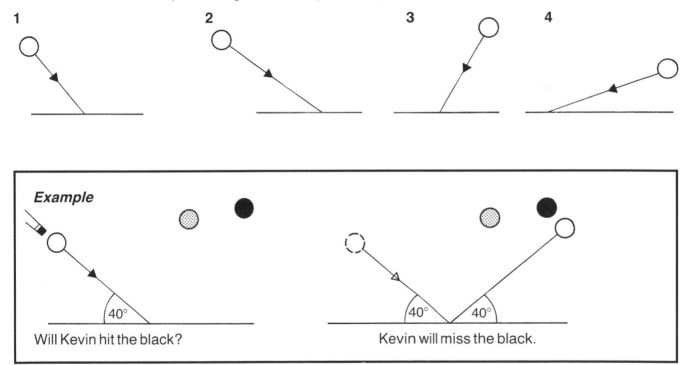

Example

Will Kevin hit the black?

Kevin will miss the black.

Trace each diagram. Find out whether Kevin will hit the black or not.

5   **6**   **7**

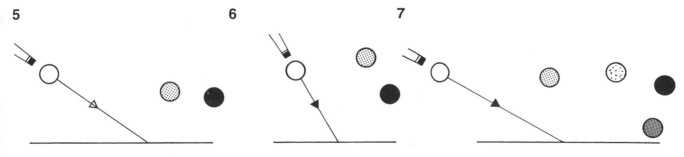

High and wide

Scale drawings are used to calculate heights and distances you cannot measure.
Here are some scale drawings from the class field trip.

Example
How wide is the river?

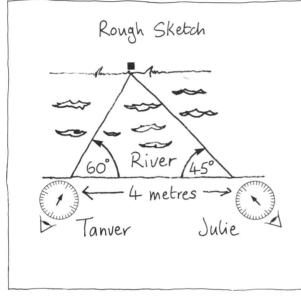

Rough Sketch

River
60° 45°
← 4 metres →
Tanver   Julie

Draw a scale drawing by following the steps.
Do you get the same answer?

Step 1
Choose a suitable scale.

cm   1   2   3   4

Scale: 1 cm represents 1 m

Step 2
Draw the scale diagram.

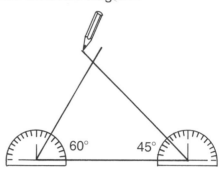

60°   45°

Step 3
Measure the width from the scale diagram.

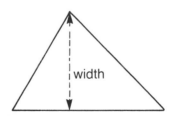

width

Width measures 2.6 centimetres.
Width of real river is 2.6 metres.

Solve each of these problems using a scale drawing.

1 Find the
height of
the tree.

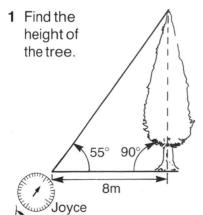

55° 90°

8m

Joyce

2 Find the width
of another river.

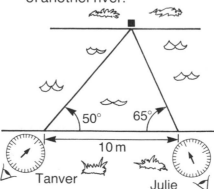

50°   65°

10 m

Tanver

Julie

3 Find the height of the
church steeple.
Use the scale
1 cm represents
2 m.

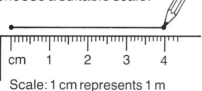

50°

20 m

Ski jumper

Example

Measure the 2 angles between the ski jumper and her skis.

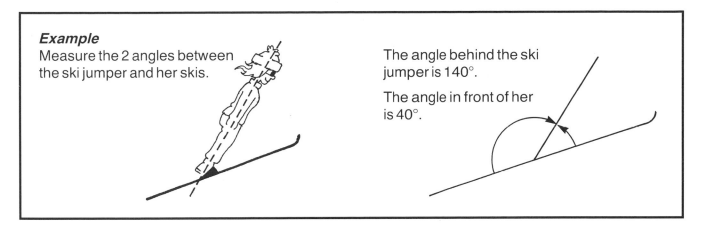

The angle behind the ski jumper is 140°.

The angle in front of her is 40°.

1 Measure both angles between the skier and her skis.

(a)   (b)   (c)   (d)

2 Write down what you notice about the 2 sizes in each answer in question 1.

Show your teacher what you have written.

3 Work out the missing size in each diagram.
Do not measure any of the angles.

Example

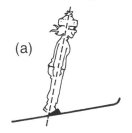

$$?° + 125° = 180°$$
$$?° = 180° - 125°$$
$$?° = 55°$$

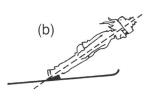

(a)

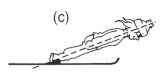

(b)

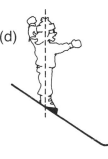

(c)

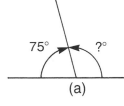

(d)

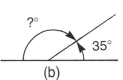

(e)

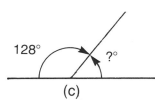

4 Work out the missing angles in this crazy paving.

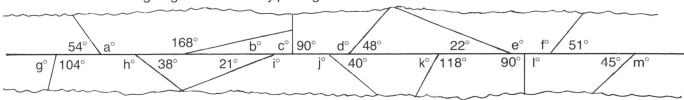

Angles in a triangle

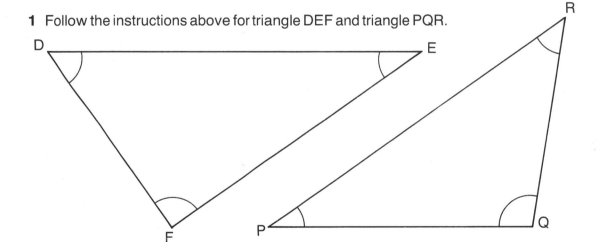

Example
Measure each angle of triangle ABC.
Write your answers in a table like this.

Angle	Size in degrees
A	80°
B	40°
C	60°
Total	?°

1 Follow the instructions above for triangle DEF and triangle PQR.

2 Copy this sentence and fill in the missing number.

The angles of a triangle add up to _____°.

T Show what you have written to your teacher.

3 Someone has torn a corner off each of the 7 triangles below.
Measure every angle then match each torn-off corner to the
correct triangle.
(Each angle is a multiple of 5°.)

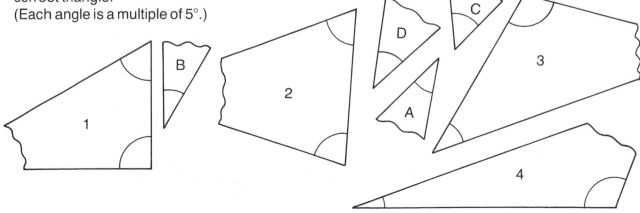

Four-sided shapes

1 Each of the pictures below shows a 4-sided shape which is part of an everyday object.

A 4-sided shape is called a **quadrilateral**.
It also has 4 angles.

Measure the angles in each shape, then put them in a table.

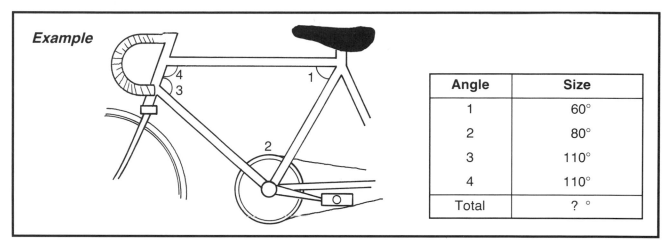

Angle	Size
1	60°
2	80°
3	110°
4	110°
Total	? °

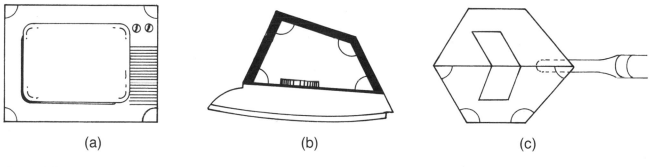

(a)   (b)   (c)

2 Copy this sentence and fill in the missing number.

The angles of a quadrilateral add up to ____°.

T Show your teacher what you have written.

3 Each of the quadrilaterals below has had a corner torn off.
By measuring all the angles, match up the pieces which make complete quadrilaterals.

This quadrilateral has 2 corners missing.

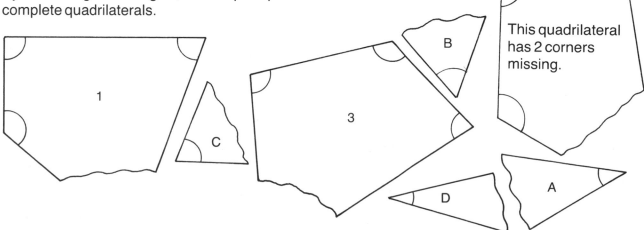

38

Full circle

1 (a) Trace each of these triangles on to tracing paper.

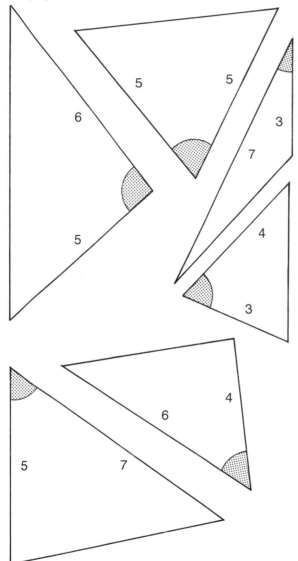

(b) Place your tracing over a piece of card. Stick a pin through the corners of each triangle to transfer them on to the card. Cut out the 6 card triangles.

(c) Write all the sizes on your card triangles (including the size of each shaded angle).

(d) Fit the triangles together by matching the 2 sides which are 3 cm long, the 2 sides which are 6 cm long, etc. and placing the shaded angles next to each other.

(e) Add up the 6 shaded angles. Explain why the shaded angles make up a complete circle.

2 Lesley cut up 3 cakes for her birthday party. All the pieces are shown below. Measure the angle in each piece, then say which pieces made up each whole cake.

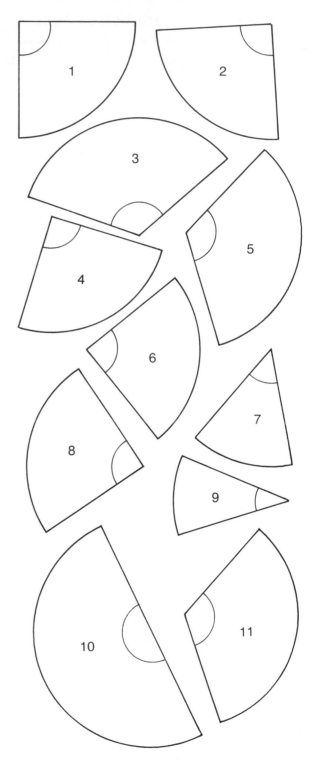

Making patterns

Some shapes are made by dividing a circle into a number of equal parts.

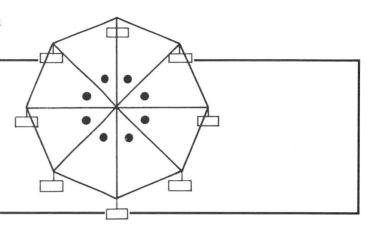

> **Example**
> This fairground big wheel is in the shape
> of a regular octagon (an 8-sided shape).
> The 8 angles marked ● total 360°.
> So each angle marked ●
> measures 360° ÷ 8
> $\qquad$ = 45°

1 (a) Draw a circle with a radius of 6 cm.
$\quad$ (b) Inside your circle draw a regular octagon.
$\qquad$ (Divide the circle into 45° sections.)

2 This diagram shows a wheel in the shape
of a regular decagon (a 10-sided shape).

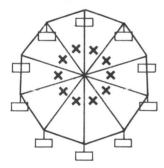

How big is each angle marked x?
Draw a regular decagon inside a circle with
a radius of 6 cm.

3 This diagram shows a regular
duodecahedron (a 12-sided shape).

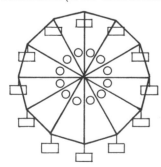

How big is each angle marked o?
Draw a regular duodecahedron inside
a circle with a radius of 6 cm.

4 Use your answers to questions 1, 2 and 3 to
help you copy each of the patterns below.
(Start from the first diagram each time.)

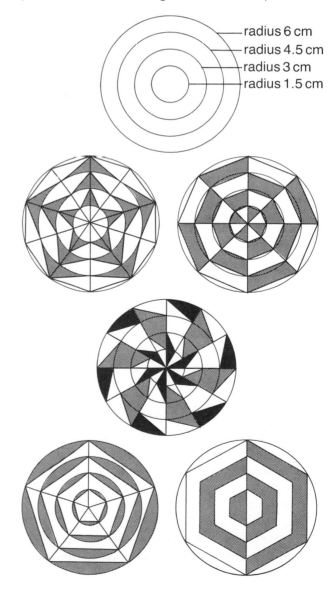

radius 6 cm
radius 4.5 cm
radius 3 cm
radius 1.5 cm

Initials

Each dot on the graph opposite stands for one student's name. ▶

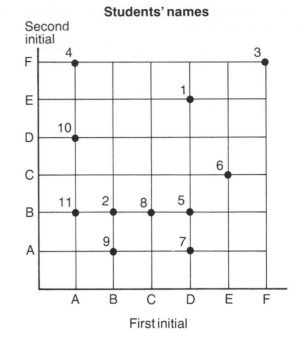

Students' names

Examples
Dot 1 stands for Daniel Evans.
Dot 9 stands for Brian Adams.

---

1 Which dot stands for Erica Clark?

2 Which dot stands for Davinder Bedi?

3 Which dot stands for Fiona Findlay?

4 Which of the pupils in the picture below does not have his or her name marked with a dot on the graph?

Astrid Davidson

Ben Bansal

Clive Best

Alison Ford

Dawn Carter

Alan Bell

Doona Ashton

Frank Dobbs

5 Make a copy of the grid above.
On your grid mark each of these names with a dot. (Put the initials next to each dot.)
Edward Anderson
Bonnie Carroll
Carla Farini

6 What size would you have to make the grid so that all possible sets of initials could be marked on it? (Only first and last names.)

7 Extend your grid so that the following names can be marked on it.
Colin Harvie
Julie King
Imram Jaffri
Mark each of these names with a dot on your extended grid. (Put the initials next to each dot.)

Setting up shop

Marie has set up her own shop, selling different teas and coffees.
This graph shows Marie's sales for the first 8 weeks of this year.

Marie's Tea Shop

Answer these questions from the graph.

1 How much did Marie sell in week 8?

2 What were her sales in week 4?

3 In which week were sales highest?
What were the sales that week?

4 In which week did sales reach exactly £400?

5 For 2 weeks running sales were the same.
Which 2 weeks were these?

6 In which other 2 weeks were sales the same?

7 This table shows Marie's sales for the next 5 weeks.

Week	9	10	11	12	13
Sales	£400	£600	£600	£300	£100

Draw a graph like the one above to show Marie's sales from week 9 to week 13.

8 Study the 2 sales graphs showing the first 13 weeks of the year.
Write down whether you think Marie is happy or upset at the way sales are going.

Triangles

Adele measured the sides of some equilateral triangles. (An equilateral triangle has 3 equal sides.)
She then drew this graph.

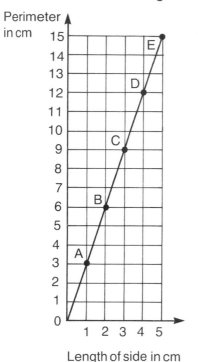

Perimeters of triangles

Length of side in cm

Example
Each side of this triangle measures 2 cm.

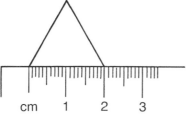

cm   1   2   3

The perimeter of the triangle is
2 cm + 2 cm + 2 cm
which is 6 cm.
Point B on the graph stands for this triangle.

1 Which point on the graph stands for this triangle?

---

2 Which point on the graph stands for this triangle?

3 One of the triangles Adele measured had each side 5 cm long.
Which point on the graph stands for this triangle?
What is its perimeter?
Draw this triangle.

4 Copy Adele's graph but extend it horizontally to read up to 8 cm. Extend it vertically to read up to 24 cm.
Join up points A, B, C, D and E, and extend the graph right to the edge of the grid.

Answer these questions from your extended graph.

5 Mark with a ● the point on your graph which stands for this triangle.
What is its perimeter?

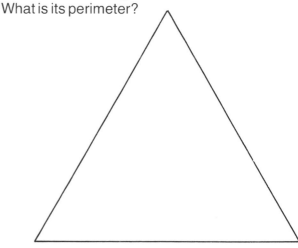

6 An equilateral triangle has a perimeter of 18 cm.
What length is each of its sides?

7 Each side of a triangle is $5\frac{1}{2}$ cm long.
What is the perimeter of the triangle?

8 A triangle has a perimeter of $22\frac{1}{2}$ cm.
What length is each side?

Springtime

Lee and her friend did an experiment to measure how far different masses stretch a spring. The results of the experiment are shown in the table below.

Mass on spring (grams)	50	100	150	200
Length of spring (mm)	30	36	42	48

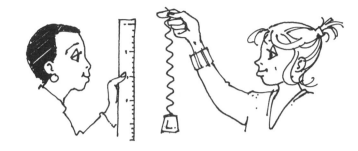

Lee then drew a graph of the results.

1 Draw the graph yourself using the scales shown below.
Your graph should go up to 60 mm and 300 g.
When you join up the points on your graph make the line go right across the grid.

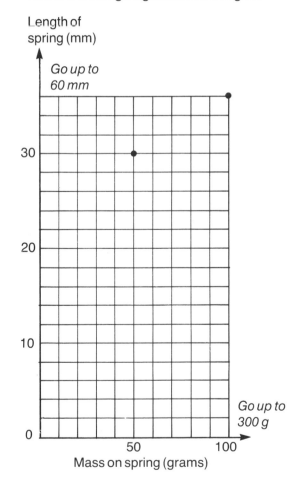

Length of spring (mm)

Go up to 60 mm

Go up to 300 g

Mass on spring (grams)

Answer these questions from your graph.

2 What length will the spring be when stretched by a mass of
(a) 250 g   (b) 300 g
(c) 170 g   (d) 220 g?

3 What mass is hanging on the spring when its length is
(a) 40 mm   (b) 50 mm
(c) 35 mm   (d) 58 mm?

4 What length is the spring with no mass attached to it?

5 What mass doubles the length of the spring compared to its unstretched length?

6 Copy and complete this sentence.

Each extra 50 g stretches the spring by
. . . . . . mm.

7 Lee and her friend repeated the experiment with a stronger spring.
Here is what they discovered about this second spring.

Spring 2

Unstretched length 30 cm.
Each extra 50g stretches spring by 4mm.

Draw a graph for the second spring like the one for the first spring.

Sporting speeds

1 The world record for running 100 metres is
about 10 seconds.
Imagine the world's fastest runner could keep
up this speed for ever.

> ### Example
> The 400 m race is 4 times as long as the
> 100 m, so the super athlete would take
> 4 × 10 seconds, that is 40 seconds.

(a) Copy this table and fill in the 'superhuman'
times for each race.

Length of race (metres)	100	200	400	800	1000	1500	2000	3000	5000
Superhuman time (seconds)	10		40						
World record (mins: secs)	0:10	0:20	0:44	1:42	2:12	3:31	4:51	7:32	13:00
World record (seconds)	10			102					

(b) Change each world record time into
seconds and fill in the bottom line of the
table. ▶

> ### Example
> Time for 800 m is 1:42, which is
> 60 seconds + 42 seconds or
> 102 seconds.

(c) Draw graphs to show the 'superhuman'
times and the actual world records on
2 mm squared paper.
Use the scales shown opposite.
Your graph should go up to 800 seconds
on the time axis and 5000 metres on the
length axis.

Join up each set of points carefully.
Explain the difference between the 2
graphs.

(d) From your graphs, write down the
'superhuman' time and the world record
time for
  (i) the 4000 m   (ii) the 2500 m
  (iii) the 10 000 m

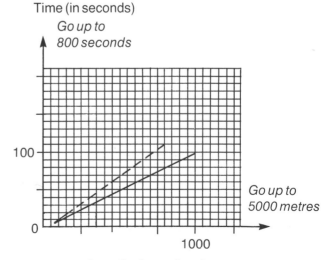

Time (in seconds)
*Go up to
800 seconds*

100

0

1000

*Go up to
5000 metres*

Length of race (in m)

2 Here are some finishers in the London Marathon, together with their finishing times.

Paul	Colin	Alice	Bert

$2\frac{1}{2}$ hours $7\frac{1}{2}$ hours   5 hours   10 hours

(a) Copy and complete the table below.

> **Example**
> Alice took twice as long as Paul.
> She must have run at *half* his speed.
> Alice's speed is $\frac{1}{2}$ of 10 mph which is 5 mph.

	Paul	**Alice**	**Colin**	**Bert**
Time	$2\frac{1}{2}$ hours			
Average speed	10 mph			

(b) Draw a graph to show all the times and speeds.
Use the scales shown below.
Your graph should go up to 10 mph on the speed axis and 10 hours on the time axis.

Join up the points on the graph with a smooth curve.

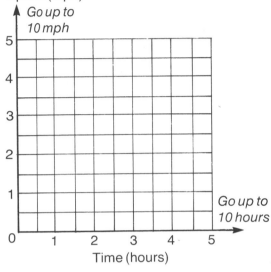

Use your graph to answer these questions.

(c) Tanya took $8\frac{1}{2}$ hours to finish.
What was her approximate average speed?

(d) John managed to keep up an average speed of 7 mph.
Roughly how long did it take him to finish the course?

46

Thinking things out

Many of these problems can be solved using a number line.

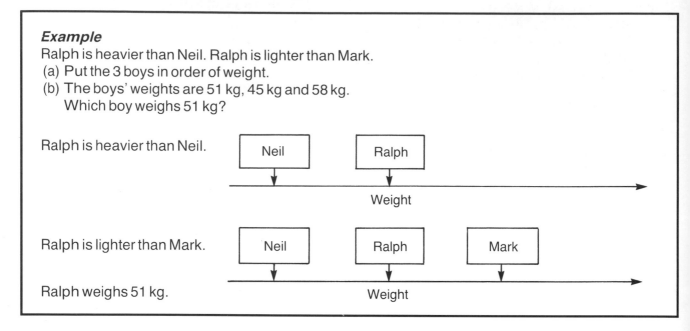

Example
Ralph is heavier than Neil. Ralph is lighter than Mark.
(a) Put the 3 boys in order of weight.
(b) The boys' weights are 51 kg, 45 kg and 58 kg.
    Which boy weighs 51 kg?

Ralph is heavier than Neil.

Ralph is lighter than Mark.

Ralph weighs 51 kg.

1 In a class vote, *All my love* was a more popular song than
   Starlight. *Fast Lady* was less popular than *Airship*, and *Starlight*
   was more popular than *Airship*.
   (a) Put the 4 songs in order of popularity.
   (b) Copy the graph and write the name of each song under the
       correct column. ▶

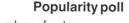

Popularity poll
Number of votes

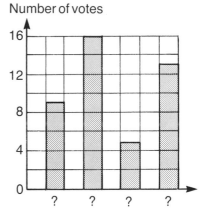

2 The Echo is dearer than the News. The Herald is dearer than the
   News. The Herald is cheaper than the Echo.
   (a) Put the 3 newspapers in order of cost.
   (b) The 3 papers cost 17p, 20p and 22p. Which paper costs 20p?

3 Copy this grid which contains only one letter G. ▶
   Fill in the letters on the grid using these clues.
   G is under E. G is to the right of A. A is below R. D is to the right of R.
   N is under A. W is to the right of O. G is above O. E is to the right of G.

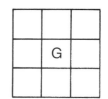

4 In snooker the 7 colours are worth from 1 point to 7 points.
   Pink is worth double the green, yellow is worth double the red,
   and brown is worth double the yellow. Black is worth more than
   blue.
   Draw and colour the 7 balls, in order, with the correct number of
   points written above each.
   You must be able to explain how you solved this one.

5 Last week, Monday was colder than Thursday which was colder than Tuesday. Wednesday was warmer than Thursday but colder than Tuesday. Friday was exactly the same temperature as Wednesday.
   (a) Arrange the days in order of temperature.
   (b) One of these graphs correctly shows last week's temperatures.
   Copy the correct graph.

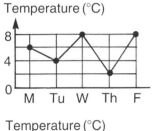

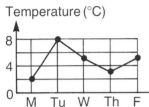

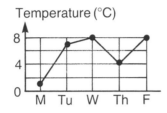

6 The 4 masses shown in these sketches are 2 kg, 3 kg and 5 kg.
   Which is which?

   You must be able to explain to your teacher how you solved this problem.

7 In these additions the figures 1 to 6 have been replaced by different symbols.

```
  ●  7        ●  9        △  ✳
+ ●  ▲      + ✳  □      + ■  8
─────       ─────       ─────
  8  9        8  0        1  ▲  1
```

Extra clue
   ■ is greater than △
   (a) Work out the meaning of each symbol.
   (b) Work out the answer to this addition.

```
   ✳  △  □
+  ▲  ■  ●
─────────
```

8 Mark scored less than Asha and Lynne in the maths test.
   Mark beat Robert who got the same score as Stuart.
   Jane scored less than Robert. Penny beat Mark.
   Each of these 7 pupils' marks is shown on the graph.
   Make a list of names showing each pupil's correct score.

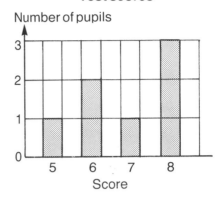

Test scores

9 This diagram shows a classroom seating plan.

```
Row B  [    ][    ]   [    ][    ]   [ J ][    ]

Row A  [    ][    ]   [    ][    ]   [    ][    ]
         1    2         3    4         5    6
```

Example
Joe sits in seat (5B).
(See J on plan.)

   (a) Copy the plan. (Make the desks big enough for you to write names in.)
   (b) Fill in each student's name using the information below.
       Tom sits in seat (5A).
       Joe, Aneela and Salim sit in odd-numbered seats in the same row.
       Salim's number is bigger than Aneela's.
       The next 4 students all sit in the same row as Tom.
       Peter's number is $\frac{1}{2}$ of Jason's number, and Jason's is $\frac{1}{2}$ of Patrick's.
       Clive's number is less than Patrick's but greater than Jason's.
   (c) Which seats are empty?

Covering up

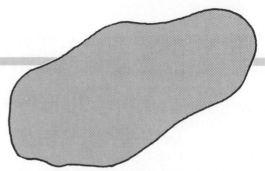

T Ask your teacher for the tracings or photocopies of the six tilings.
What is the area of this shape?
You can use your tilings to measure the area.

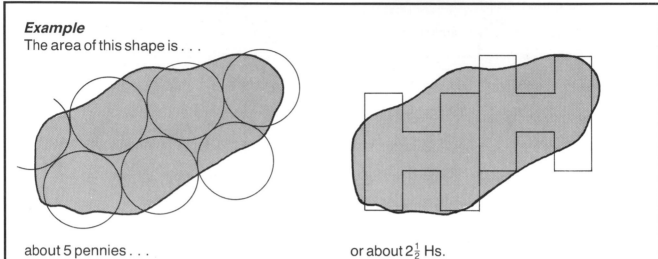

Example
The area of this shape is . . .

about 5 pennies . . .   or about $2\frac{1}{2}$ Hs.

Copy and complete the following sentences.

1

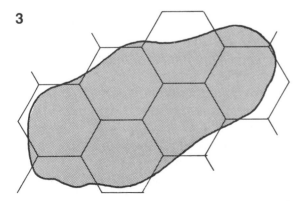

The area is about — swans.

2

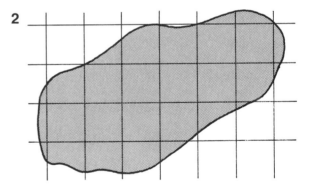

The area is about — squares.

3

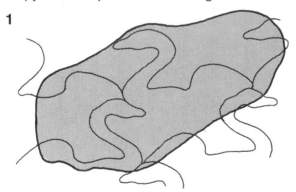

The area is about — hexagons.

4

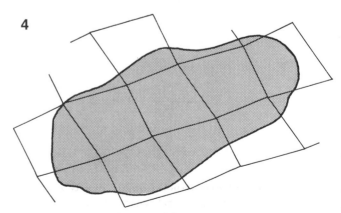

The area is about — kites.

Find the area of each shape in pennies, Hs, hexagons, swans, squares and kites.

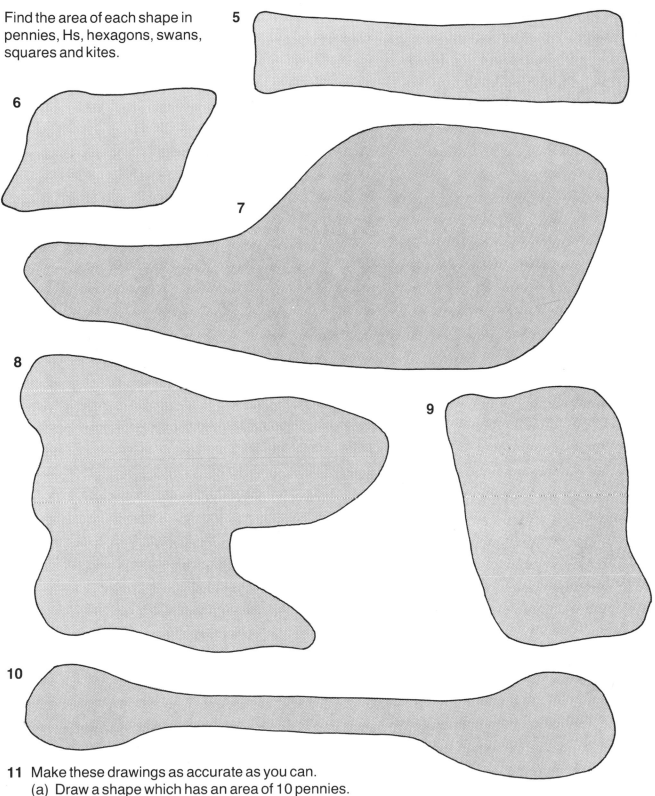

5

6

7

8

9

10

11 Make these drawings as accurate as you can.
   (a) Draw a shape which has an area of 10 pennies.
   (b) Draw a shape which has an area of 5 swans.
   (c) Draw a shape which has an area of 8 hexagons.
   (d) Draw a shape which has an area of 3 Hs.
   (e) Draw a shape which has an area of 9 kites.
   (f) Draw a shape which has an area of 20 squares.
   Show the tiles on each drawing to prove that your answers are
   about right.

Shapes and sizes

1 Try to put these shapes in order of area (from smallest to biggest) by looking at them.

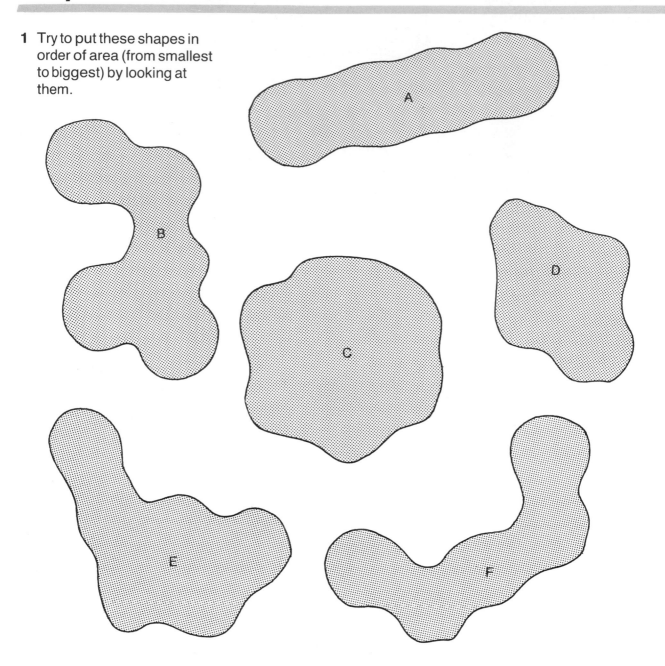

2 Now find the area of each shape in *hexagons*. Put the shapes in order. Were you right the first time?

3 Draw two different shapes whose area is one hexagon less than the smallest shape.

4 Draw two different shapes whose area is one hexagon more than the largest shape.

5 Shape G has about the same area as one of the shapes above. Which one?

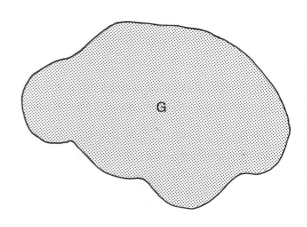

6 Try to put these shapes in order of area (from smallest to biggest) by looking at them.

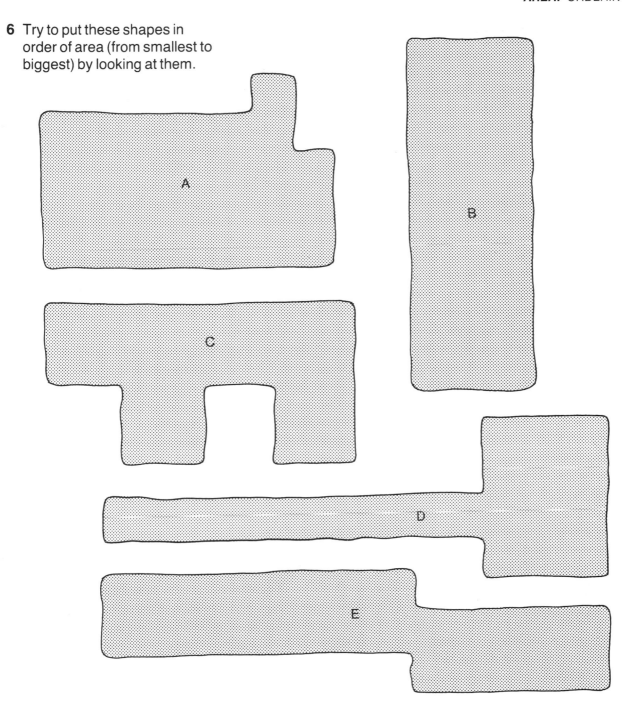

Now find the area of each shape in *squares*.
Put the shapes in order. Were you right the first time?

7 Draw two different shapes whose area is one square less than the smallest shape.

8 Draw two different shapes whose area is one square more than the largest shape.

9 Shape F has about the same area as one of the above shapes. Which one?

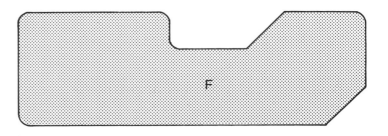

Square centimetres

These three shapes all have an area of 1 cm².
Write down in cm² the shaded area of each shape
below.

Square metres, square yards

How big is 1 m²?
Make it out of paper.
Look at the example opposite. ▶

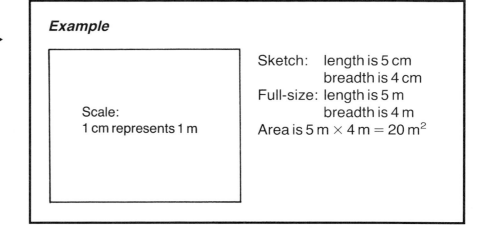

Example

Scale:
1 cm represents 1 m

Sketch: length is 5 cm
        breadth is 4 cm
Full-size: length is 5 m
        breadth is 4 m
Area is 5 m × 4 m = 20 m²

Work out the area of each
full-size rectangle in m². ▼

1

Scale:
1 cm represents 2 m

2

Scale:
1 cm represents 1 m

3

Scale:
1 cm represents
1 m

4 How big is 1 yd²?
Make it out of paper.
Work out the area of this
rectangle in yd².

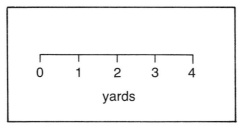

0   1   2   3   4

yards

Work out the area of each
shape in ft².

5

0   1   2   3

feet

6

0   1   2   3

feet

Feed the lawn

1 Mr Brown and Mr Smith agree to buy a packet
of Lawncare between them.
This is a scale drawing of their lawns.

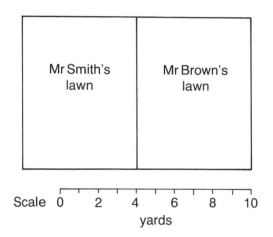

Scale 0 2 4 6 8 10
yards

LAWNCARE
100 SQ. YDS.
NORMALLY
£5.99
OUR PRICE
£4.99

(a) Will one packet of Lawncare be
enough?
Write down all the working you need to
find out.

(b) How much should they each pay?

(c) Draw a sketch of a square lawn which would need
exactly one whole packet of Lawncare.
Use the same scale as above.

(d) Draw two different rectangular lawns which
would need a whole packet of Lawncare.

---

2 Mr Potts, a gardener, wants to treat this lawn
with Lawncare.

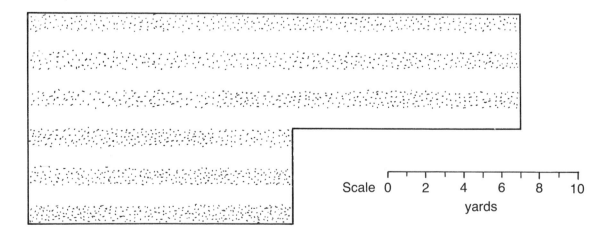

Scale 0 2 4 6 8 10
yards

(a) What is the area of the lawn?

(b) How many packets of Lawncare must
he buy?

(c) What is the cost of Lawncare per square
yard (to the nearest penny)?

(d) Mr Potts uses the entire contents of all the
packets he bought on the lawn.
How much did this cost per square yard
(to the nearest penny)?

3 The Willow family and the Hay family decide to buy some Weedkil and some Evergrow between them.
The plan opposite shows the lawns to be treated. ▶

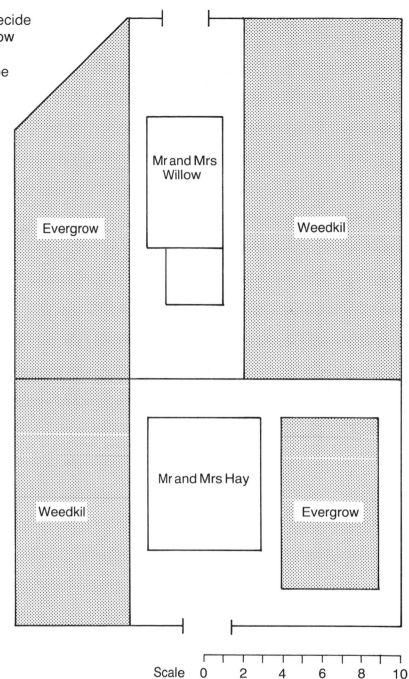

Scale 0 2 4 6 8 10
metres

(a) Work out the total area to be treated with each fertiliser.
Writing your answers in this table will help you.

	Weedkil	Evergrow	Total
Willow	152 m²	96 m²	
Hay			
Total			

(b) How many bags of each fertiliser should they buy?
(c) How much will this cost?
(d) How much should the Hay family pay as their share of the bill:
£2 – £3, £5 – £6, £8 – £9, £12 – £13 or some other amount?
Explain your choice. (Use your table from part (a)).

64

Length, breadth and perimeter

T 1 With a piece of string 20 cm long, try and make the six rectangles in the table. ▶
Copy and complete the table.

	Length (cm)	Breadth (cm)	Perimeter (cm)	Area (cm²)
1	4	6	20	24
2		3	20	
3	8			
4				25
5	3			
6	6			

2 The three graphs show the rectangles you can make from a piece of string 20 cm long.
Each ■ covers a number.
The numbers covered are **10**, **20** and **25**.
Each ▬▬▬ covers a label.
The labels covered are '**Area (cm²)**', '**Breadth (cm)**' and '**Perimeter (cm)**'.
See if you can fill in the missing labels.
Note These graphs are not to scale.

(i) Find out which graphs, numbers and labels go together.
(ii) Draw each graph accurately.

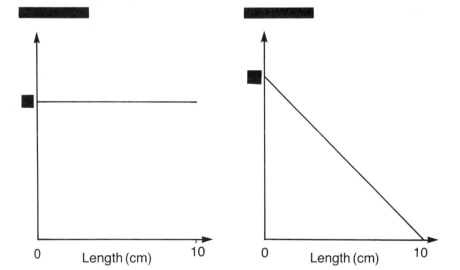
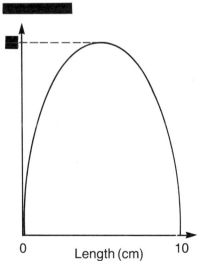

Use the correct graph to answer these questions.
Each question refers to a rectangle made from 20 cm of string.

3 What length is a rectangle of breadth 2 cm?

4 What area is a rectangle of length 8 cm?

5 What is the perimeter of a rectangle of length 7 cm?

6 What breadth is a rectangle of length 5 cm?

7 What is the biggest area of any of the rectangles?

8 What length is the rectangle with the biggest area?
What is special about this rectangle?

An area investigation

Example

Think about these questions:

Is it possible to make a square with an area of 36 cm²?

If so, what length is the side of the square?

How many rectangles can you make which have an area of 36 cm² (whole numbers only)?

What are the dimensions of these rectangles (whole numbers only)?

This table gives the answers to these questions for areas of 36 cm² and 8 cm².

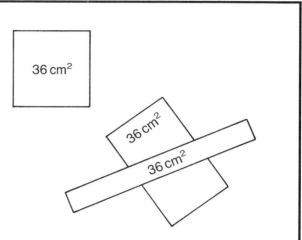

Area	Square	Rectangles
36 cm²	6 cm × 6 cm	1 cm × 36 cm, 2 cm × 18 cm, 3 cm × 12 cm, 4 cm × 9 cm
8 cm²	None	1 cm × 8 cm, 2 cm × 4 cm

1 Make out a table which answers the 4 questions for every area from 1 cm² to 100 cm². ▶

2 Areas which can be made into squares are called square numbers.
Make a list of all the square numbers between 1 and 100.

Area	Square	Rectangles
1 cm²	1 cm × 1 cm	None
2 cm²	None	1 cm × 2 cm
3 cm²	None	1 cm × 3 cm
100 cm²		

3 Work out the length a.

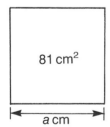

81 cm²

a cm

4 Work out the length b.

16 cm²

b cm

5 Work out the lengths x and y.

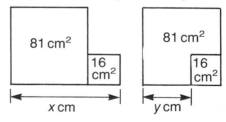

81 cm² 16 cm²

x cm

81 cm² 16 cm²

y cm

6 Work out the length l.

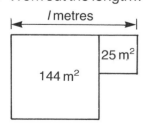

l metres

25 m²

144 m²

7 Work out the length x.

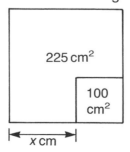

225 cm²

100 cm²

x cm

8 Work out the length l.

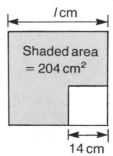

l cm

Shaded area = 204 cm²

14 cm

Volume

1 You are going to find out how many centimetre cubes container A can hold.

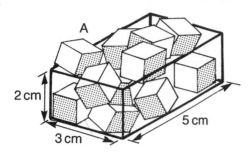

(a) Why is this not a very accurate way of solving the problem?

(b) If you fitted the cubes in carefully, how many would cover the bottom of the container?
This picture will help you.

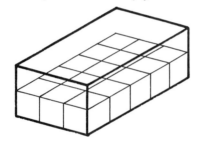

(c) How many layers of cubes like this would the container hold?
This picture will help you.

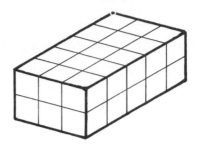

(d) How many cubes does the container hold altogether?

(e) Work out the answer to this calculation:
$5 \times 3 \times 2$.

(f) What do you notice about your answers to parts (d) and (e)?

We say that the volume of this container is 30 cubic centimetres or 30 cm³.

---

2 Copy this table, and add 4 more rows.

Container	Volume by counting cubes	Length L cm	Breadth B cm	Height H cm	$L \times B \times H$ = Volume
A	30 cm³	5 cm	3 cm	2 cm	$5 \times 3 \times 2 = 30$ cm³

Fill in your table for each of the shapes below.

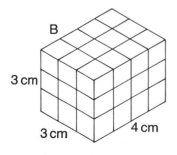

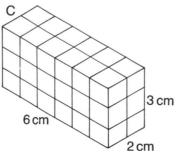

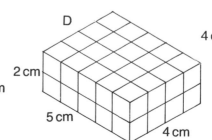

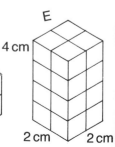

3 Work out the volume in cm³ of each of the boxes below.

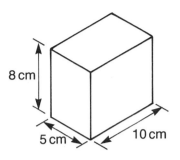

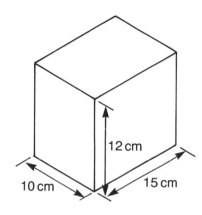

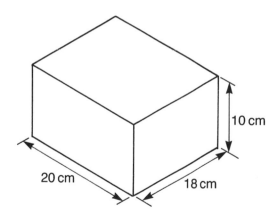

4 (a) Work out the volume of each box in the table.

Box	Length	Breadth	Height
a	14 cm	8 cm	5 cm
b	13 cm	10 cm	4 cm
c	8 cm	8 cm	8 cm
d	9 cm	11 cm	6 cm

(b) Arrange the boxes in order of volume, from smallest to largest.

5 Small tins of cocoa are filled from the large tin.

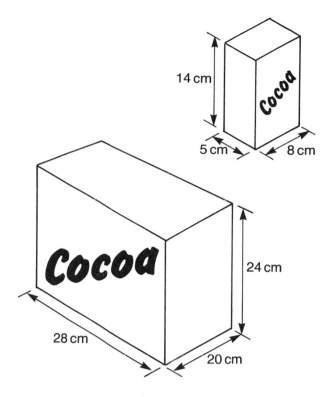

(a) Work out the volume of the small and the large tin.

(b) How many of the small tins can be filled from one large tin?

T 6 Your teacher will give you some boxes. Measure the length, breadth and height of each box and then work out its volume.

7 Larger volumes are measured in cubic metres (m³). Work out the volume of the room shown in the sketch. Your answer will be in m³.

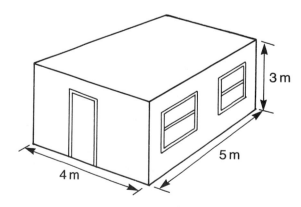

8 (a) Work out the volume of your classroom.

(b) Work out the volume of your living room at home.

Filling up

The volume of a liquid is usually measured in **litres**.
1 litre = 1000 cm³

Example 1
How many litres of water can this tank hold?

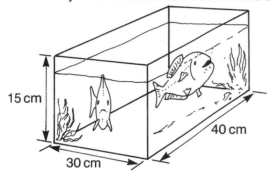

Volume is   40 cm × 30 cm × 15 cm
= 18 000 cm³
= 18 litres

Example 2
Paula fills the tank to a depth of 12 cm.
How many litres of water are in the tank?

Volume is   40 cm × 30 cm × 12 cm
= 14 400 cm³
= 14.4 litres

---

1 (a) How many cm³ of water does this fish tank hold when filled to the brim?

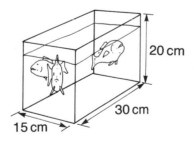

(b) How many litres is this?
(c) How many litres of water are in the tank when Jason fills it to a depth of
   (i) 10 cm
   (ii) 12 cm
   (iii) 8 cm?
(d) Jason pours 6.3 litres of water into the empty tank.
   How deep is the water?

2 (a) What special name is given to a container this shape?

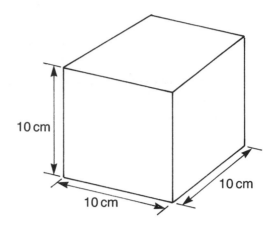

(b) How much liquid would it hold?
(c) This would be an unusual shape for a milk carton. Work out the dimensions of other containers which would hold 1 litre.
(Try doubling one of the sides and halving another)
(d) Make every carton you design. Write 'I litre' on each one.

3 Design a can to hold 5 litres of petrol.

4

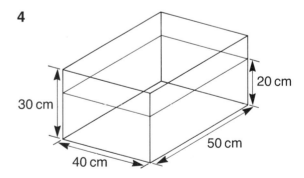

(a) Mark fills this tank to a depth of 20 cm. How much water is in the tank?

(b) Make a scale model of the tank. (Use the scale 1 cm represents 5 cm) Draw a line on the outside of your model to show the water level.

(c) Mark puts a block measuring 25 cm by 20 cm by 20 cm into the tank.
Work out the new water level.

(d) Draw the new water level on your model.

5 You have 3 empty bottles like this.

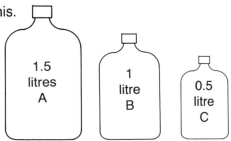

(i) Write down exactly what you would do to solve each of these problems.

(ii) Show a calculation to prove that your solution is correct.

| 1 litre = 1000 ml |
| 0.5 litre = 500 ml |

Experiment	Problem
(a) Fill bottle B with water. Bottle C is empty.	How many times can you fill bottle C from bottle B?
(b) Fill bottle A with water. Bottle C is empty.	How many times can you fill bottle C from bottle A?
(c) Fill bottle B with water. Bottle A and bottle C are empty.	Pour out water until you know you have exactly half of the water left.
(d) Fill bottle A with water. B and C are both empty.	Pour water from A into B until B is full. How much water is left in A?
(e) Fill bottle A with water. B and C are both empty.	Pour some water out so that there is exactly 1 litre left in bottle A.
(f) Fill bottle A with water. B and C are both empty.	Pour water from bottle A until all 3 bottles contain exactly the same amount of water.
(g) Fill bottle A and bottle C with water. Bottle B is empty.	Pour water from A and C until 2 of the bottles contain equal amounts of water and the third bottle has twice as much as the other two.

6 You have 3 empty lemonade bottles. They are labelled 1 litre, 750 ml, and 500 ml. You also have another large empty bottle. Explain how you would pour these amounts of water into the large bottle:

(a) $1\frac{1}{2}$ litres

(b) $1\frac{3}{4}$ litres

(c) $\frac{1}{4}$ litre

(d) $1\frac{1}{4}$ litres

Saturday afternoon

Mark and Syeda went to watch Manchester United play Everton.

Study these clues carefully.

Clue 1 Mark and Syeda support Manchester United.

Clue 2 The half-time interval lasts a quarter of an hour.

Clue 3 At half-time they decided they could eat only half a hamburger each.

Clue 4 A football match has $1\frac{1}{2}$ hours playing time.

Clue 5 Everton scored 10 minutes after half-time.

Clue 6 Five minutes of injury time were added to the second half today.

Clue 7 It took them half an hour to get to Mark's house.

Clue 8 The game started at exactly 3.00 pm today.

Clue 9 Manchester United scored 5 minutes before half-time.

Clue 10 Manchester United play in red shirts.

Answer these questions. The clues will help you.
1 At what time was the first goal scored?
2 How much did Mark and Syeda spend on hamburgers?
3 At what time did they join the queue for hamburgers?
4 At what time did the full-time whistle go?
5 What was the final score?

6 At what time did they get to Mark's house?
7 Who is the captain of Manchester United?
8 What is Manchester United's ground called?
9 Which of the clues did you not need to help you answer the questions?
10 Which questions cannot be answered from the clues?

Mark's mum and his sister Viv found this recipe in an old book.
All the packets of flour and sugar in the kitchen are marked in kilograms.
The kitchen scales are marked in kg and g.

PERKINS

4 oz. plain flour	$\frac{1}{2}$ tsp. cinnamon
2 oz. butter or margarine	$\frac{1}{2}$ tsp. ground ginger
4 oz. fine oatmeal	$\frac{1}{4}$ tsp. mixed spice
3 oz. castor sugar	3 level tbsps. syrup
1 level tsp. baking soda	1 oz. blanched almonds

   Prepare a greased baking sheet. Sieve flour and rub in the butter. Mix in all the other ingredients except the almonds and syrup. Mix to a firm dough with the syrup. Spoon pieces the size of a walnut on to baking sheet, leaving room to spread, and flatten them with a fork. Place half an almond on each. Bake in a moderate oven (350° to 375°) for 10 minutes, or until crisp. Cool on a wire rack. **Makes about 15 biscuits.**

11 Do what Viv's mum asked *her* to do.
Write out the list of ingredients with the weight of each item in grams.

12 Viv's mum put the mixture in the oven at about a quarter to two.
When were the biscuits ready?

Viv and her mum went out to the cinema (a 10 minute walk away) in time for the start of the Saturday matinee.

13 How much did they pay altogether to get in?

14 The first film on the programme lasted one hour and the second film lasted an hour and a half. There was an interval of a quarter of an hour between films.
At what time did they leave the cinema?

On their way home they noticed this advert in the TV showroom window.

15 Viv worked out how much they would save altogether by hiring this video.
How much was it?
They spent about half an hour discussing this before deciding not to take it.

16 Who arrived home first, Mark and Syeda or Viv and her mum?
How much of a gap was there between them?

Astoria

Daily 7.00 pm
Matinees 2.30 pm
Adults £2.50
Children $\frac{1}{2}$ price if accompanied by adult.

Video Hire
$\frac{1}{2}$ price rental

£7.49 per month for the first 2 months
Normal rental £14.99 per month

72

Half time

T 1 (a) Lift a 1 kg packet of sugar.
   (b) Your teacher will give
       you 4 or 5 objects.
       One of these weighs $\frac{1}{2}$ kg.
       Which one?
       Lift each object to find out.
   (c) Check your answer
       using scales.

T 2 (a) Look at a metre stick.
   (b) Your teacher will give
       you 4 or 5 sticks.
       One of these is $\frac{1}{2}$ metre long.
       Which one?
       Look at the sticks to find out.
   (c) Check your answer using
       the metre stick.

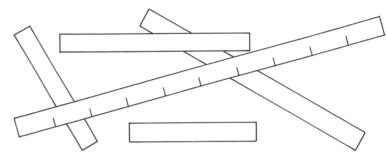

T 3 (a) Look at a one litre bottle.
   (b) Your teacher will give you
       4 or 5 containers.
       One of these holds $\frac{1}{2}$ litre.
       Which one?
       Look at the containers to
       find out.
   (c) Check your answer with
       your teacher.

T 4 (a) You are going to try to
       estimate a $\frac{1}{2}$ minute.
       You need a friend with a
       stopwatch. Your friend will
       tell you when to start.

       You must wait for a $\frac{1}{2}$
       minute and then shout
       'stop'. Your friend will tell
       you the reading on the
       stopwatch when you
       shouted 'Stop!'
       If the reading is 30 seconds,
       you are spot on!
   (b) Repeat this 5 times. Work out
       your average answer (rounded
       to one decimal place).
   (c) The table shows Roberta and
       Mark's estimates (in seconds).

Your friend

Go!

You

Stop!

	Roberta	Mark
1	38.35	25.42
2	28.29	29.63
3	29.59	30.60
4	27.76	29.38
5	27.54	31.79
Total	151.53	146.82
Average	30.3	29.4

5 Copy this rectangle.
See how many different ways
you can shade one half of it.

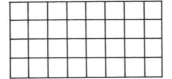

Three possibilities
are shown to
start you off.

6 Here are two of Jean's
answers to question 5.

Answer 1

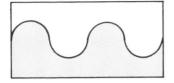

Answer 2

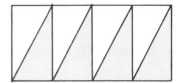

Pete is making the comments.
What would you say to Pete?

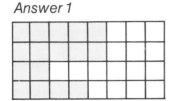

Never! The two
parts are not the
same shape.

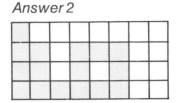

This can't be right.
The shaded bit is
not the same
shape as the other
bit . . . *and* it's got a
hole in it!

7 Write down your comments on these tea-time conversations

Tuesday tea-time

We'll have half of
this cake each.

That's not fair!
Your half is bigger
than mine!

Thursday tea-time

There are three of
us tonight. I'll just
have to cut the
cake into three
halves.

74

8 Copy and trace this shape.

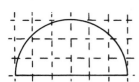

Each of the following shapes is made by giving the half circle a half turn.
Find out where to put the pin for each one.
Draw each shape.

(a)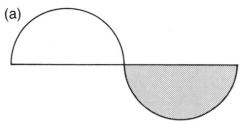

(b)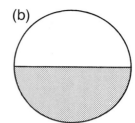

(c)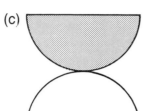

(d)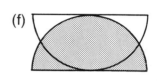

(e)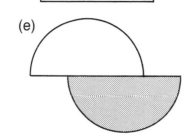

(f)

9 Repeat for this shape.

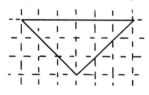

(a)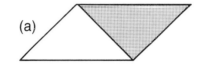

(b)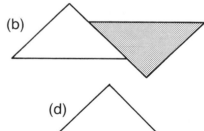

(c)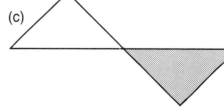

(d)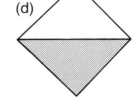

10 Do as before for this shape. The dividing line between the two halves has been missed out this time.

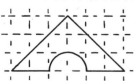

(a)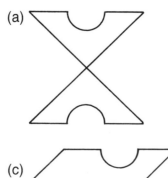

(b)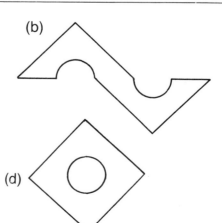

(c)

(d)

Ice champions

The finalists in the ice dancing world championships came from Canada, Great Britain, USA, West Germany and Russia.
The table below shows the scores given by the judges to each of the first 4 couples.

	Canada	GB	USA	W. Ger.
Judge A	5.6	6	5.8	5.7
Judge B	5.7	6	5.8	5.7
Judge C	5.7	5.9	5.7	5.7
Judge D	5.5	6	5.4	5.4
Judge E	5.4	5.9	5.6	5.6
Judge F	5.7	6	5.6	5.7
Judge G	5.8	6	5.7	5.6
Total points				

1 (a) Work out each couple's total score.
  (b) What is the medal position so far?

2 Dave has no calculator.
  He says that Great Britain is 4th so far with a total of 14.8!

```
        6
        6
      5·9
        6
      5·9
        6
        6
      ────
      14·8
```

  (a) How did he get this answer?
  (b) How would you explain to him what he has done wrong?

3 The Russian couple was last to skate.
  Here are their scores:

Judge	A	B	C	D	E	F	G
Score	5.8	5.7	5.7	5.6	5.4	5.6	▪

Now look at the medal ceremony.

How many points did Judge G give the Russian skaters?
Explain how you got your answer.

4 What scores from judge G would have given the Russians 2nd place?

Sharpen up

1 Some of these readings are correct and some are not. Write down the letter of each correct reading.
Now rearrange the letters you have written down to spell the name of something with a point (not a decimal one!).

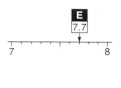

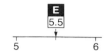

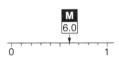

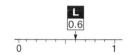

2 (a) Copy this grid, but *don't* copy the numbers.
 (b) Replace each number by its letter from one of the seven scales below.
 Hint You will find an arrow pointing to 5.9 on Scale 1.
 (c) Write the letter in the blank square where the number was.
 (d) Keep going until all the blanks in your grid are filled with letters.
 (e) If you read the grid in the correct order you will find the names of three more things with a point.
 (f) Write down the three names.

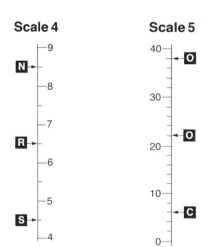

Scale 1

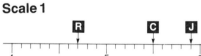

Scale 2

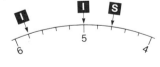

Scale 3

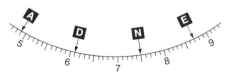

Scale 4

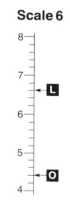

Scale 5

Scale 6

Scale 7

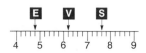

Back to school

Davinder bought these things the day before going back to school.
The assistant in the shop charged him £5.80. She used the cash register to get the total.

When Davinder got home he checked his bill using a calculator.
This is what he did:

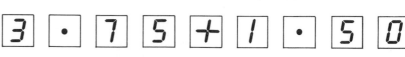

He asked his sister Kulvinder to check it again.
This is what she did:

Who is right and who is wrong?
Explain the mistakes that were made.

Measure them all

Look at the pictures on this page and the next.
Answer these questions correct to the nearest 0.1 cm.

1 What is the length of
   (a) the pair of scissors
   (b) the comb
   (c) the pencil
   (d) the middle finger
   (e) the nail on the forefinger?

2 What are the length and breadth of
   (a) the post card
   (b) the stamp?

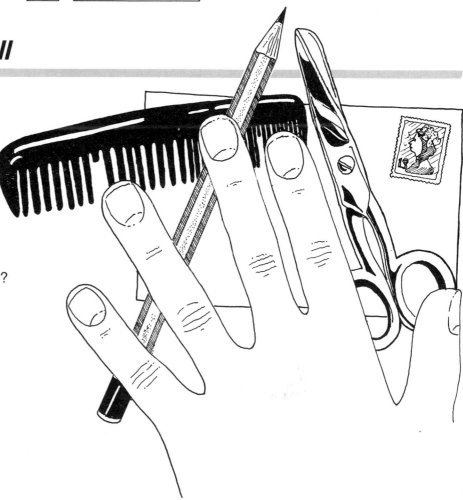

3 What is the length of
(a) the key
(b) the needle
(c) the pencil in the compasses?

4 What is the diameter of the 10p?

5 (a) What is the diameter of the circle the compasses are set to draw?
(b) To draw a circle the same size as the 10p, how far apart would you set the compass needle and pencil?

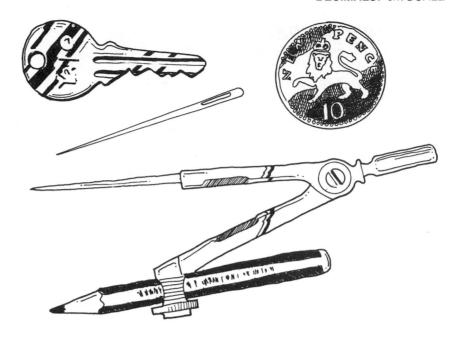

6 Measure the dimensions marked a, b and c in the drawings of a plug.

7 (a) What is the width of the plug?
(b) An extension socket box takes three plugs side by side. There must be at least 2 cm between plugs and 2 cm at each end of the box.
What length must the box be at least?

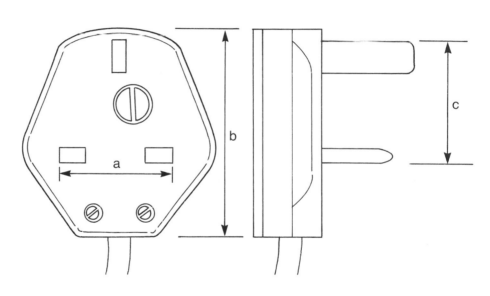

Answer each of the questions below to the nearest 0.1 cm.

8 What is the diameter of the outer circle?

9 What is the perimeter of the triangle?

10 What is the perimeter of the square?

11 What is the diameter of the inner circle?

12 What is the length of the diagonal of the square?

13 What is the perimeter of the regular hexagon?

14 What is the perimeter of the rectangle?

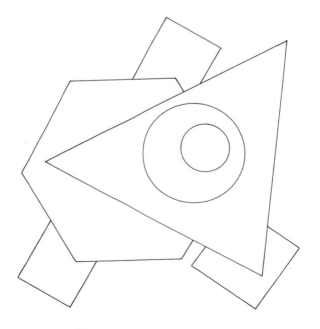

90

Battery power

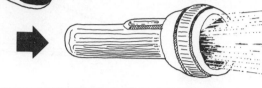

This is an HP11 battery.
(1.5 volts means the same as $1\frac{1}{2}$ volts.)

Example
A torch needs 2 HP11 batteries. This makes a voltage of 3 volts.

---

1 A radio cassette recorder needs 4 HP11 batteries.
What voltage is this?

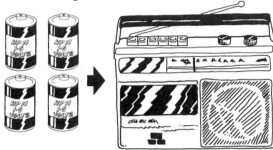

2 A portable stereo needs 6 HP11 batteries. How many volts is this?

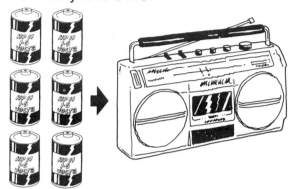

3 Draw a chart to show the total voltage of different numbers of HP11 batteries, from 1 battery up to 8 batteries.

Here is how to start your chart:

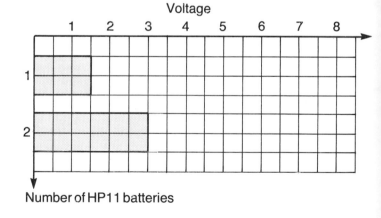

4 Now use your chart to answer these questions:
   (a) What voltage is made by 6 HP11 batteries?
   (b) What voltage is made by 8 HP11 batteries?
   (c) A radio needs 9 volts to make it work.
       How many HP11 batteries does it need?
   (d) A torch needs 6 volts.
       How many HP11 batteries does it need?
   (e) A tape recorder needs 7.5 volts.
       How many HP11 batteries are required?
   (f) Mrs Todd bought three 4.5 volt torches for her children.
       How many HP11 batteries does she need to buy?
   (g) An HP11 battery costs 55p. Jenny spent £2.75 on batteries for a radio. What is the voltage of the radio?
   (h) How many volts are made by 10 HP11 batteries.
       Explain how you worked out your answer.
   (i) A large radio needs 18 volts to function.
       How many HP11 batteries will be needed?
       Write down at least two different ways of working out this answer.

Tee time

This is a full-size sketch of a golf ball.
Design a box to hold 3 of these golf balls.
Make a scale drawing of the net of your box.
Use the scale 1:2.

Draw a full-size net on card and paste up the box.

Reminder
This diagram reminds you what the net of a simple box looks like.

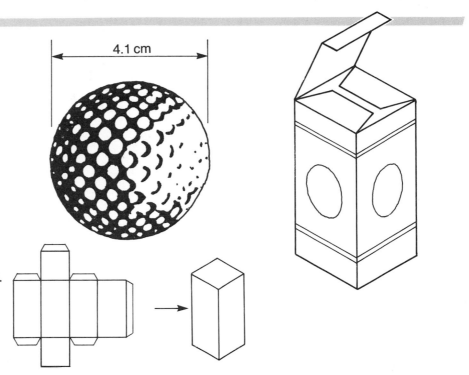

4.1 cm

Power puzzle

These two grids give you a code for the different letters.

Use these grids to crack the codes below. Find five articles which might run off batteries. Your teacher will give you a clue if you cannot get started.

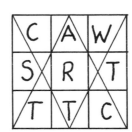

M	E	C
C	⊙	U
H	A	L

C	A	W
S	R	T
T	T	C

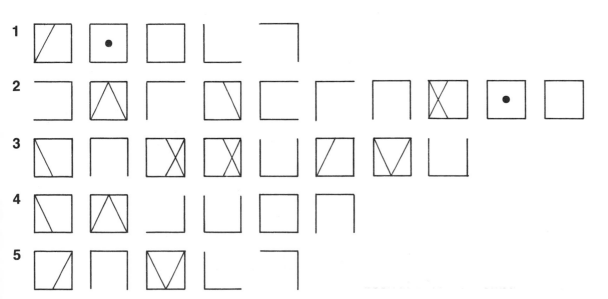

1

2

3

4

5

Number drill 1

Check these multiplications on a calculator:
Did you notice how the calculator display changed each time?

1 $52.7 \times 10 = 527$ **2** $4.5 \times 10 = 45$

3 $14.75 \times 10 = 147.5$ **4** $0.56 \times 10 = 5.6$

Do these on your calculator. Notice how the display changes. Write down the calculations as above.

5 7.8×10	**6** 19.9×10	**7** 0.47×10
8 3.64×10	**9** 0.8×10	**10** 1.05×10

Do these without using your calculator.

11 8.4×10	**12** 15.2×10	**13** 3.65×10
14 0.75×10	**15** 10×4.7	**16** 10×18.06
17 3.25×10	**18** 0.09×10	**19** 10×17.83

---

20 Find the cost of 10 of each of these articles.
The cost of one is shown on the price tag.

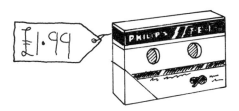

21 Work out the total weight of 10 of these parcels.

22 Here is what Joe said in his answer to question 21. How would you explain his mistake to him?

To multiply any number by 10 you just put a 0 on the end like this . . .
$10 \times 7 = 70$
$10 \times 4 = 40$
See?
So $10 \times 7.6 \text{ kg} = 7.60 \text{ kg}$.
10 parcels would weigh 7.60 kg.

23 A car does 35.7 miles to the gallon.
How far should it travel on 10 gallons of petrol?

24 In a factory making plastic cups, a cup comes off the assembly line every 1.6 seconds.
How long does it take to collect a set of 10 cups?

25 Dareeta buys ten 1.5 litre bottles of lemonade.
How many litres is this?

Voltage values

Here is a formula used for electrical circuits.

> Voltage = Current × Resistance

Voltage is measured in **volts**.
Current is measured in **amps**.
Resistance is measured in **ohms**.
Work out the voltage in each of these circuits.

1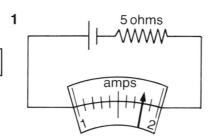

5 ohms

amps

2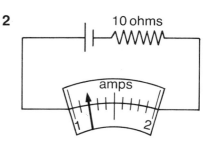

10 ohms

amps

3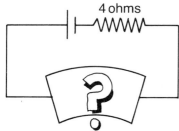

$7\frac{1}{2}$ ohms

amps

4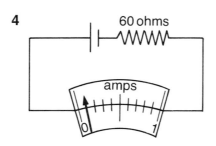

60 ohms

amps

5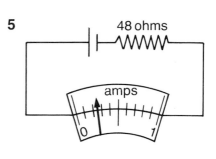

48 ohms

amps

6 This circuit has a 6 volt battery.
Draw a scale to show the current reading.

4 ohms

7 This circuit has a 12 volt battery.
What is the resistance?

?

amps

8 This circuit has a 6 volt battery.
What is the resistance?

?

amps

9 This graph shows currents and resistances in a 12 volt circuit.
Answer the questions from the graph.
  (a) The current in a circuit is 3 amps.
      What is the resistance?
  (b) The current is 2 amps.
      What is the resistance?
  (c) The resistance in a circuit is 20 ohms.
      What is the current?
  (d) The resistance is 3 ohms.
      What is the current?

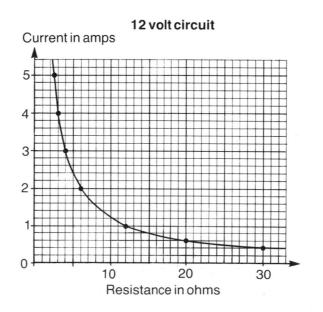

12 volt circuit

Current in amps

Resistance in ohms

94

Rounding

Example

Work out the deposit on this sofa (to the nearest penny).

Deposit = $\frac{1}{3}$ of £500

£500

$\frac{1}{3}$ DEPOSIT

Working:

| 5 | 0 | 0 | ÷ | 3 | = | *166.66666* |

166.66

166.67

The deposit is either £166.66 or £166.67.
But 166.66666 is nearer 166.67.
The deposit is £166.67.

---

1 Joe, Karen and Rehana share the prize.
How much should each get (to the nearest penny)?
Explain why the answer to this question is not the same as the answer to the example.
How much of the prize cannot be shared?

GRAND PRIZE DRAW
£500
(and not one penny more)

2

$\frac{1}{3}$ *DEPOSIT*

Work out the deposit on each of these (to the nearest penny).

Cassette .... cash price £139
Camera .... cash price £59.99
Cutlery .... cash price £42.50
Video .... cash price £449
Hi-fi .... cash price £348

3 Divide each sum of money into three equal shares (to the nearest penny).
(a) £139   (b) £42.50
(c) £59.99   (d) £449
How much of each sum is left over?

---

4 Seven factory cleaners won £57 000.
How much should each receive?
Do this on your calculator:

| 5 | 7 | 0 | 0 | 0 | ÷ | 7 | = | |

Why does nobody mention pennies this time?

£8000 each
for seven cleaners

---

5 Copy and complete these tables:

Answer from calculator	Rounded to nearest penny	Rounded to nearest pound
£45.3683	£45.37	£45
£37.1835		
£345.7194		
£107.09724		
£39.86281		

Answer from calculator	Rounded to two decimal places	Rounded to one decimal place
45.87327 m	45.87 m	45.9 m
1.5832 litres		
15.70923 kg		
22.3475 cm		
119.8635°		

Number drill 2

Use a calculator

1 4.75×10	**2** 4.75×100	**3** 4.75×1000	**4** 47.5×100
5 675×10	**6** 675×100	**7** 67.5×10	**8** 6.75×1000

Do not use a calculator

9 3.14×10	**10** 3.14×100	**11** 3.14×1000	**12** 31.4×100
13 496×10	**14** 496×100	**15** 49.6×10	**16** 49.6×100
17 4.68×10	**18** 4.68×100	**19** 4.68×1000	**20** 46.8×100
21 9.07×10	**22** 9.07×100	**23** 9.07×1000	**24** 90.7×100

Use a calculator

25 0.76×10	**26** 7.6×10	**27** 76×10	**28** 760×10
29 0.58×100	**30** 5.8×100	**31** 58×100	**32** 580×100

Do not use a calculator

33 0.85×10	**34** 8.5×10	**35** 85×10	**36** 850×10
37 0.96×100	**38** 9.6×100	**39** 96×100	**40** 960×100

Use a calculator

41 $570 \div 10$	**42** $57 \div 10$	**43** $5.7 \div 10$	**44** $57 \div 100$
45 $380 \div 10$	**46** $38 \div 10$	**47** $3.8 \div 100$	**48** $3.8 \div 10$

Do not use a calculator

49 $540 \div 10$	**50** $54 \div 10$	**51** $5.4 \div 10$	**52** $54 \div 100$
53 $750 \div 10$	**54** $75 \div 100$	**55** $7.5 \div 100$	**56** $75 \div 10$
57 $6150 \div 10$	**58** $615 \div 100$	**59** $61.5 \div 100$	**60** $615 \div 10$

Work out the answers mentally:

61 One shirt costs £6.75. What do 10 shirts cost?

62 One magazine costs £0.85. What do 100 magazines cost?

63 Ten tickets cost £54.80. What does one ticket cost?

64 One pin costs £0.07. What do 100 pins cost?

65 Ten televisions cost £575. What does one TV cost?

66 100 kg potatoes cost £47. What is the cost per kg?

67 10 bricks stretch 2.2 m. How long is one brick?

68 One bottle holds 1.5 litres. How much do 1000 bottles hold?

69 1000 photocopies cost £35.70. What does one cost (in pence)?

Example
Ms King earns £6500 per annum. How much is this per week?

Yearly salary is £6500
Weekly wage is £6500 ÷ 52
= £125

---

70 Miss Smith earns £7592 per annum. How much is this per week?

71 Mrs Davidson's salary is £4966 per year. What is her weekly wage?

72 James Granger's annual salary is £9126. How much is this per week?

73 Gloria Glitter earned £24 973 last year. How much is this per week?

74 Mr Ali's annual salary is £15 000. How much is this per week?

75 How much per week is an annual allowance of £494?

Tithing

Many religious groups tithe their income.
This means that each member gives one-tenth of his or her wages
to the church every week.
How much does each of these church members give each week?

1 Julie Campbell, joiner
£85 per week

2 Peter Donaldson, plumber
£78 per week

3 Nita Suleman, teacher
£6240 per annum

4 John Forbes, OAP
£34.50 per week

5 Helen Davies, doctor
£15 860 per annum

6 Chu Yu Wong, schoolgirl
£5 per week

7 Ronald Hutchison,
unemployed
£22 per week

8 Jane Brown, civil servant
£612 every four weeks

9 Paul Lewis, decorator
£175 per week

10 Sue Evans is a teacher earning £10 140 per
annum.
She gets a rise which brings her salary up to
£10 660 per annum.
How much more per week does she give to
the church after her rise?

11 Maria gives £18.50 to the church each week.
What is her annual salary?

Play your cards right

Make a set of 10 cards like this:
You can now set out these
cards to make a sum.

In the following questions you
should use all three numbers
and any of the signs you wish.

1 Set out your cards for this calculation:
Find the answer using your calculator.
Copy down the calculation and the
answer.

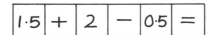

Repeat for these calculations. Always do the part in brackets first.

2
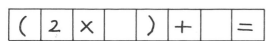

3
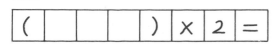

4
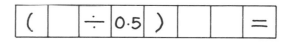

5
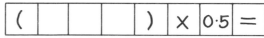

6 Set out the cards to make the answer come
to 4.
Hint This can be done by changing one card
in the calculation in question 4.
Copy down the calculation.

7 Set out the cards to make the answer come
to 0.

8 Make the answer 3.5.
Hint

$(\boxed{2} \times \boxed{}) + \boxed{} =$

9 Make the answer 2.
Hint

$(\boxed{} \boxed{} \boxed{}) \times \boxed{2} =$

10 Make the answer 0.5.

11 Make the answer 6.

12 Make the answer 5.
Hint

$(\boxed{} \div \boxed{0.5}) \boxed{} \boxed{} =$

13 Make the answer 0.25.
Hint

$(\boxed{} \boxed{} \boxed{}) \times \boxed{0.5} =$

14 Make up some calculations of your own.
Ask your friend to solve them.

15 What is the biggest answer you can make?

16 What is the smallest answer you can make
(apart from 0)?
No negative numbers allowed.

17 How would you make the answer 1?

18 The $\boxed{0.5}$ card is replaced by another
number.

$(\boxed{1.5} + \boxed{?}) \times \boxed{2} = \boxed{}$ 11

What is the other number?

19 The $\boxed{0.5}$ card is replaced by another
number.

$(\boxed{2} - \boxed{1.5}) \div \boxed{?} = \boxed{}$ 0·5

What is the other number?

A circle formula

This formula tells us how to calculate the circumference of a circle:

$$C = \pi \times D$$

C stands for the circumference of the circle.
π is always 3.14.
D stands for the diameter of the circle.

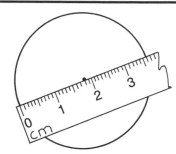

Example
Calculate the circumference of this circle.
Give the answer correct to two decimal places.

$C = \pi \times D$ $\pi = 3.14$
$ = 3.14 \times 3.6$ $D = 3.6\,cm$
$ = 11.304$
$ \simeq 11.3$
Circumference of circle is 11.3 cm.

---

1 Calculate the circumference of each of these circles.
Give your answers correct to one decimal place.
Set down your working as in the example.

      (a)   (b)   (c)

2 This is a scale drawing of a penny-farthing bicycle.
Scale 1:20
Calculate the circumference of each of the wheels of the full-size machine.

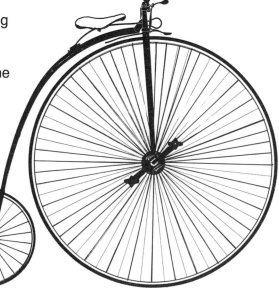

3 Calculate the total length of the lines drawn to make this pattern.
Work out the circumference of one circle first, then count up the number of circles.
It may help if you copy the pattern.

Hexacut

The shapes on this page were made by cutting this hexagon into pieces and fitting the pieces together again.

Work out which pieces were used to make each shape.

Make good drawings to show how you solved each problem. Your teacher will give you hints if you get stuck.

1

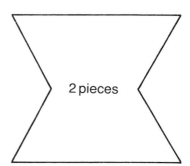

2 pieces

2

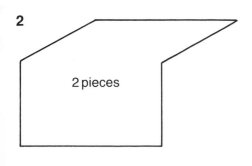

2 pieces

3

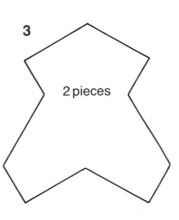

2 pieces

4

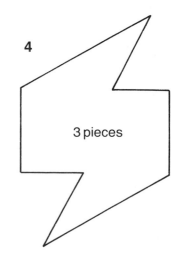

3 pieces

Problems with a point: quick questions

1 Mary won the 100 metres race at the school sports in a time of 13.5 seconds. Karen finished 2.6 seconds behind Mary. What was Karen's time for the race?

2 Joe is 1.58 metres tall. Pete is 1.71 metres tall. How much taller is Pete than Joe?

3 Sue is on a 10 km sponsored walk. After 6.5 km she feels like giving up. What did her friend say to encourage her?

Come on, Sue! You've only got ___ km to go!

4 How far could this car travel on 5 gallons of petrol?

BOCKLEY SPECIAL 42.8 mpg

5 This snooker triangle was made from a piece of wood 46.5 cm long.

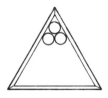

What length is each side of the triangle?
The wood was cut like this:

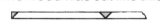

6 What would happen if you tried to empty the two smaller bottles into the large bottle?
  (a) Would there be a space in the large bottle or would it overflow?
  (b) What volume would the space or overflow have?

7 (a) What is the cost of five 165 SR 13 tyres at Treadeezi?
  (b) What would these tyres cost at list price?
  (c) How much do you save by buying at Treadeezi?

Treadeezi Tyres

Size	List price	Treadeezi price
155 SR 13	£41.86	**£20.90**
165 SR 13	£46.02	**£23.90**
165 SR 14	£47.70	**£25.70**
175 SR 14	£53.23	**£27.70**

8 Work out the radius of this circle.

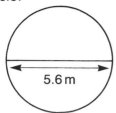

5.6 m

9 Each of these circles has a diameter of 3.2 m.
Calculate the length of the straight line.

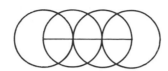

10 The radius of each circle is 1.5 m.
Calculate the perimeter of the square.

11 Joe is 1.58 m tall, Pete is 1.71 m tall, Calvin is 1.64 m tall and Halim is 1.83 m tall. What is the average height of these boys?

12 The school record for the 4 × 100 m relay race is 51.2 seconds. This year the times were: Karen 12.4s, Salima 13.1s, Tanya 12.6s and Rita 12.4s.
By how much did this team beat the record?

13 An encyclopaedia has 12 volumes. Each volume is 3.5 cm thick.
How much space does it take up on the shelf?

14 The table below is from a breakfast cereal packet.

Nutritional content per 100 g	
Energy	350 kcal
Protein	7.9 g
Iron	6.7 g
Vitamin B6	1.8 mg

How much of each of these will you get from a 25 g helping?

15 Copy and complete this order form.

Developing and processing charges			
	Price	Number	Cost
12 exposures	£1.40	2	
15 exposures	£1.85	—	
20/24 exposures	£2.25	3	
36 exposures	£2.65	1	
	Totals		

The firm also charges 25p postage per film.
What total amount must you send?

16 Calculate the weight needed to balance
A + B + C + D.

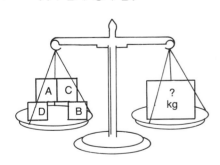

Clue 1

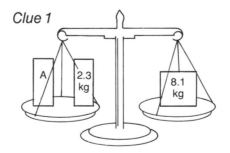

Clue 2

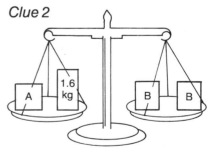

Clue 3

Clue 4

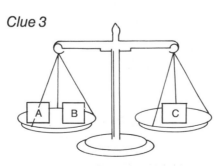

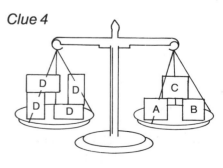

102

Round the room

Reminder
1 m = 100 cm   So we can write   1 m 75 cm as 1.75 m   and   3 m 8 cm as 3.08 m.

Write each of these measurements in metres:
1 1 m 55 cm   **2**   4 m 86 cm   **3** 5 m 60 cm   **4** 10 m 87 cm
5 5 m   8 cm   **6** 12 m   5 cm   **7** 50 cm   **8** 5 cm

9 Work out the dimensions marked
a and *b* in this sketch of a
living room.

10 A metal strip is to be fitted
round the edge of the room
to hold the carpet down.
What length of strip is
needed?
Can you see a quick way of
working out this answer?

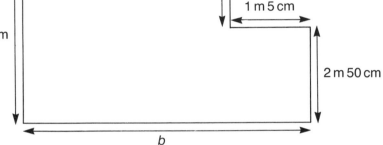

11 Use a tape measure to find the length
and breadth of your classroom.
Work out the perimeter of your
classroom in metres.

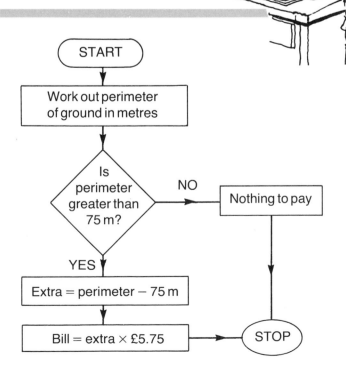

Fence me in

The local council are giving
grants to house owners for new
fencing, including gates.
The council will pay for fencing
up to 75 m per household.
House owners pay for any
extra fencing needed at the
rate of £5.75 per metre.
The rules for working out the bill
are shown in this flowchart.

The next page shows plans of
some grounds which are to
have new fences.
Work out the bill each house
owner must pay (if any).

START

Work out perimeter
of ground in metres

Is
perimeter
greater than
75 m?

NO → Nothing to pay

YES

Extra = perimeter − 75 m

Bill = extra × £5.75 → STOP

1 5 Hillend Road
Owner: Miss C Maclean

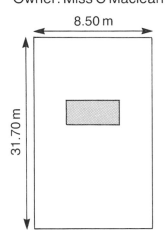

2 17 Chestnut Street
Owner: Mr J Bond

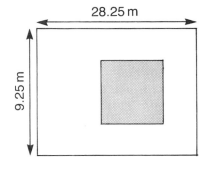

3 35 Station Road
Owner: Mrs M Cheung

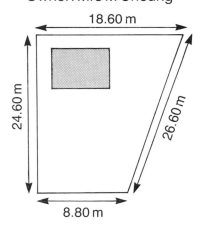

4 6 Columba Mansions
Owner: Mr C Gibb

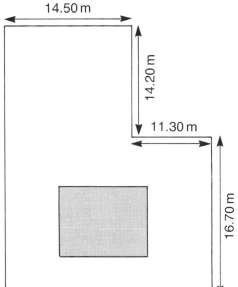

5 47 Acacia Avenue
Owner: Usha Ginda
Reminder
The circumference of any circle is roughly 3 times its diameter.
(Circumference = 3.14 × diameter gives a more accurate answer.)

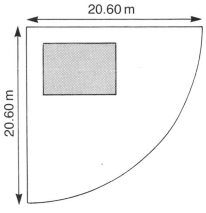

6 Only individual house owners may apply for a grant. They must submit a plan of the fencing they want. What plans should these two neighbours submit? What will each have to pay (if anything).

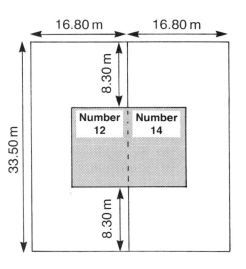

7 (a) What is the fairest way for these four householders to share the bill (if any)?

(b) Is there any way of submitting plans that would get everyone's bill paid by a grant? Would the council allow this?

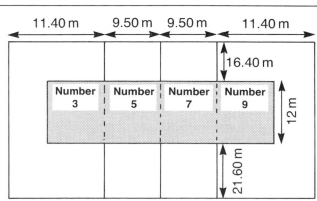

104

Upstairs

Each staircase is to be fitted with carpet 1 metre wide.
Each step is 0.25 m long and 0.15 m high.

(a) Work out the dimension W for each staircase.
(b) Work out the dimension H for each staircase.
(c) How many metres of carpet would just cover
     each staircase?

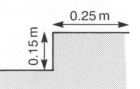

1

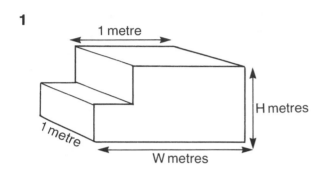

2

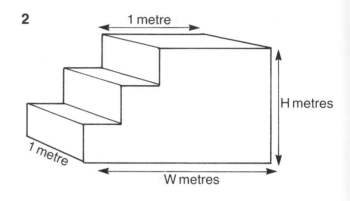

3 Draw a similar staircase with 5 steps. Mark in
all 4 dimensions: H metres, W metres, I metre,
I metre. From your diagram work out W and H.
How many metres of carpet would you need to
cover this staircase?

4 Repeat question 3 with 8 steps in the staircase.

5 This is the staircase in
Gwen's house.
Each step is the same size
as the ones in questions 1-4.
  (a) Work out dimension W.
  (b) Work out dimension H.
  (c) Gwen's mum wants a
      new carpet for the
      staircase. She has seen
      a carpet 1 metre wide
      which she likes.
      How many metres of this
      carpet will she need?
  (d) The carpet costs £15.50
      per metre.
      How much will Gwen's
      mum have to pay?
  (e) She decides to pay by
      12 monthly instalments.
      How much is each
      instalment?

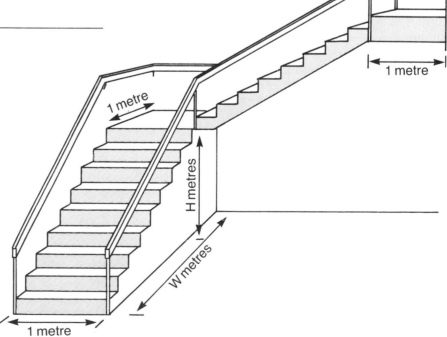

**12 months to pay
at CASH PRICE**

Falling rolls

The table below was published in 1984. It shows how the number of pupils in secondary schools is expected to change.

Numbers of secondary school pupils (thousands) 1980–1990

	North District	South District	Central District	East District	West District
1980	31.7	27.4	59.9	47.5	30.9
1981	31.7	27.3	57.6	47.0	30.2
1982	31.7	26.9	55.8	46.9	29.8
1983	31.0	26.5	53.1	45.3	29.2
1984	30.2	25.5	49.5	43.6	28.1
1985	28.9	24.5	45.9	41.5	26.9
1986	27.5	23.4	42.3	38.9	25.5
1987	26.1	22.4	39.1	36.5	23.9
1988	24.8	21.4	36.0	34.1	22.4
1989	23.6	20.6	34.1	32.1	21.2
1990	23.0	20.1	33.0	31.2	20.6

Draw a rough graph to show these figures.
The graph opposite shows you how to start.
Use a different colour for each district.

1 What is expected to happen to the number of pupils in each district?
2 The figures were worked out by the local councils. How could they predict the number of pupils in secondary schools?
3 The councils must take action based on what they have learnt from the graph. What action do you think they should take? Write down as many ideas as you can think of.
4 What could happen to change the trend of the graphs after 1990?

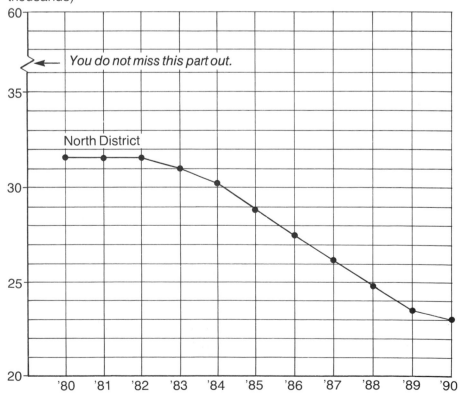

Number of pupils (in thousands)

You do not miss this part out.

North District

106

Keep taking the tablets

1 What is a prescription?

2 From the graph, what was the charge for a prescription in:
(a) 1981   (b) 1983   (c) 1984?
Give your answers correct to the nearest 10p.

3 What has happened to the cost of a prescription in the years shown on the graph?

4 Find out for yourself:
(a) what prescriptions cost now;
(b) who gets prescriptions free of charge;
(c) how the prescription charge compares with the cost of the drug.

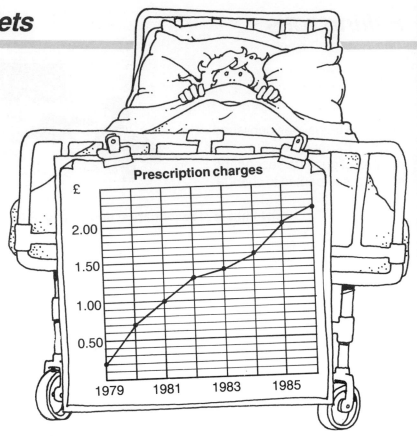

The pictogram opposite shows how much the National Health Service could save if doctors prescribed cheaper drugs. It shows, for nine different drugs, how much would be saved if doctors prescribed a cheaper version instead of the brand name.

5 What would be the savings to the NHS on drug 3?

6 One of the drugs would save the NHS £3.4m. Which drug is this?

7 Which of these drugs would save the NHS most? How much of a saving would it be?

8 What would be the total saving to the NHS on these nine drugs?

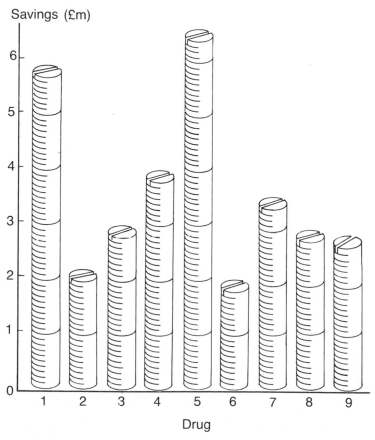

Savings to National Health Service

A magazine carried out a survey of the prices of some drugs in chemist shops.

Name	Cost of pack		Cost of 6 tablets
Paracetamol BP Brand name	82p £1.95	(100 tablets) (72 tablets)	4.9p
Aspirin BP Brand name	62p £1.49	(100 tablets) (48 tablets)	
Paracetamol/ Codeine BP	75p	(25 tablets)	
Brand name	£2.04	(24 tablets)	

9 Copy and complete the table.
Give answers to the nearest 0.1p.

10 Draw a graph to show how the price of six tablets compares for the three types of drug in the table.
Use 2 mm squared paper for your graph.

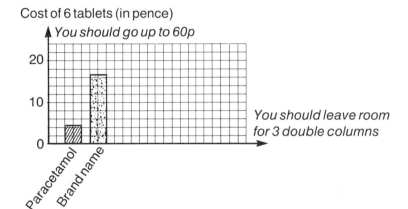

Cost of 6 tablets (in pence)

You should go up to 60p

You should leave room for 3 double columns

In the swim

This is part of a scale drawing of the side of a swimming pool. The bottom of the pool slopes evenly from one depth marking to the next. The water level is about 50 cm below the edge of the pool.

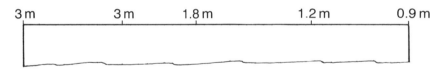

| 3 m | 3 m | 1.8 m | 1.2 m | 0.9 m |

Scale: 1 cm represents 2 m

1 What is the length of the pool?
2 How far out from the deep end do you have to go before the water becomes shallower?
Make a scale drawing to show how the bottom of the pool slopes over the whole length of the pool.
Use the scale: 1 cm represents 1 m.
Answer these questions using your scale drawing.

3 Which part of the bottom of the pool slopes most steeply?
4 If *you* started at the shallow end and waded out, how far would you get before the water went above your chin?
5 Draw yourself on the scale drawing at this place.

108

Area of a triangle

To find the area of a triangle, multiply the base of the triangle by its height, then halve the answer. Check the calculation in the example.

For each question below, measure the base and height of the triangle and then calculate the area in square centimetres. Round your answers to one decimal place.

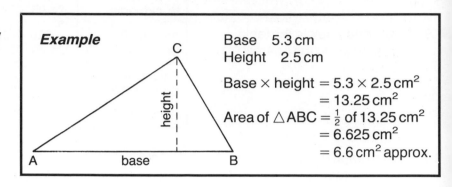

Example

Base   5.3 cm
Height   2.5 cm

Base × height $= 5.3 \times 2.5 \, \text{cm}^2$
$= 13.25 \, \text{cm}^2$
Area of $\triangle ABC = \frac{1}{2}$ of $13.25 \, \text{cm}^2$
$= 6.625 \, \text{cm}^2$
$= 6.6 \, \text{cm}^2$ approx.

1

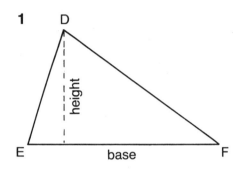

2

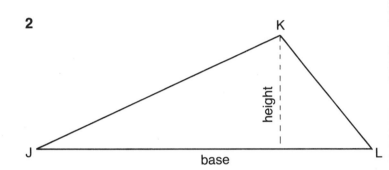

3

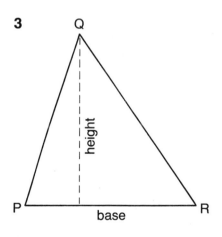

4

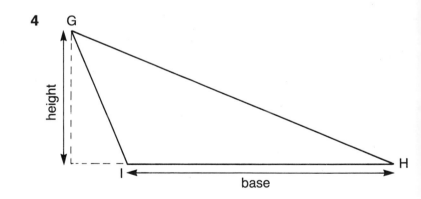

5 Divide this shape into triangles, then find its area in square metres (m²).

Hints
1 Trace the shape.
2 Scale all measurements up to metres before you do any calculations.

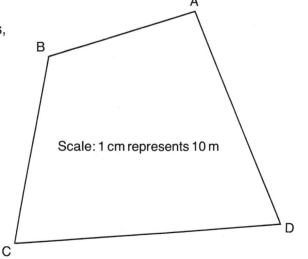

Scale: 1 cm represents 10 m

6 This scale drawing shows the gable end of a house, which is to be painted.
(a) Calculate the area (in m²) to be painted.
(b) The paint is only sold in 2.5 litre tins and 5 litre tins.
2.5 litres costs £5.25.
5 litres costs £9.75.
1 litre of paint covers 6 square metres.
Find the cost of giving the gable end two coats of paint.

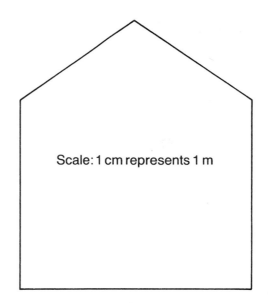

Scale: 1 cm represents 1 m

7 Calculate the area of the triangle marked on the diagram. Use this to work out the area of the whole hexagon.

8 A box 10 cm high has this hexagon as its base and top. Work out the total surface area of the box.

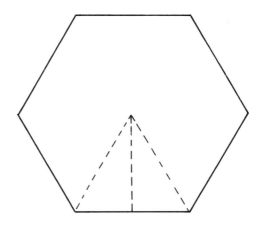

Work out the area of each of these triangles.

9

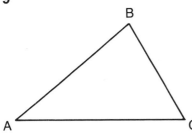

10

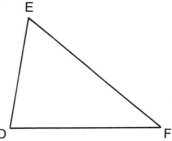

11

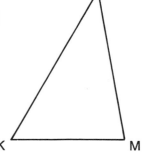

12 Why are your three answers so similar? (A tracing of the first triangle will help you.) What does this tell you about the formula on the previous page?

13 Draw a triangle which has a base of 10 cm and an area of 30 cm².

14 Draw any triangle which has an area of 45 cm². Explain how you worked it out.

Perimeters

Here are some rectangles that can be made from a piece of string 10 cm long. Measure the length, breadth and perimeter of these five rectangles.

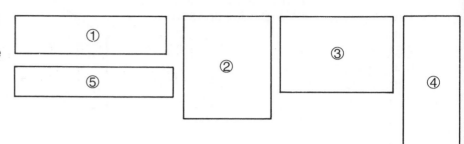

Each of these rectangles has been marked by a dot on the graph opposite. Copy the graph.

Check with your teacher that you understand how the graph works, then:

(i) Draw four more rectangles with a perimeter of 10 cm.

(ii) Mark each rectangle with a dot on your graph.

(iii) Join up all the dots on your graph with one line. (Use a ruler.)
Make the line go as far as it can on the grid.

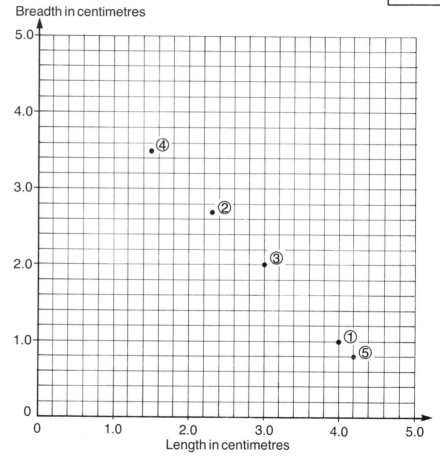

Ask your teacher to help you get started on these questions.
Use your graph to answer them.

1 The length of one of the rectangles is 3 cm.
What is its breadth?
Draw this rectangle. Check that its perimeter is 10 cm.

2 The length of one of the rectangles is 3.4 cm.
What is its breadth?
Draw this rectangle and check its perimeter.

3 The breadth of one of the rectangles is 1.6 cm.
What is its length?
Draw this rectangle and check its perimeter.
Compare your answers to question 2 and question 3.
What do you notice?

4 The breadth of one of the rectangles is 2.1 cm.
What is its length?
Draw this rectangle and check its perimeter.

5 One of the rectangles shown by the graph is a *square*.
Draw this square and check its perimeter.
How did you know where to look for the square on the graph?

6 One of the rectangles has a length of 5 cm.
What is its breadth?
Draw the rectangle.

Chubbycheeks

Chubbycheeks is a milk drink for babies.
This graph is part of an advert for Chubbycheeks. The graph compares Chubbycheeks with cow's milk.

Now answer these questions using the graph.

1 How many grams of protein are there in 100 ml of cow's milk?

2 How many grams of protein are there in 250 ml of cow's milk?

3 How many grams of fat are there in 250 ml of Chubbycheeks?

4 How many more grams of carbohydrate are there in 250 ml of cow's milk than in 250 ml of Chubbycheeks?

5 The rest of the advert describes in words the main differences between Chubbycheeks and cow's milk.
Write down what you think it says.

6 One of the pie charts opposite shows how the solid part of Chubbycheeks is made up. The other shows how the solid part of cow's milk is made up.
Which is the pie chart for Chubbycheeks and which for cow's milk?
Explain how you could tell.

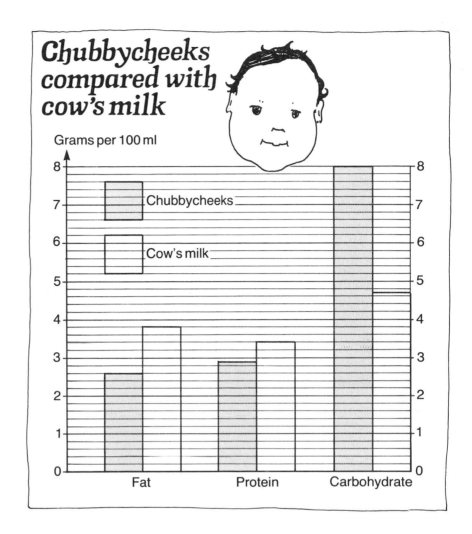

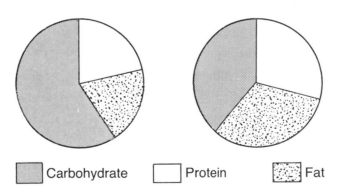

112

Paperback

This graph shows the profits for a ten-year period for Goodread, a large publishing company.

Copy the graph carefully into your notebook.

1 What were the company's profits in 1976?

2 In which years did Goodread make a loss?

3 In which year did the company make a profit of £1.1m?

4 How did the company perform in 1980? (You must have a number in your answer.)

5 By how much did the company's profits improve between 1980 and 1981?

6 In which years did the company do worse than the year before?

7 In which years did the company do better than the previous year?

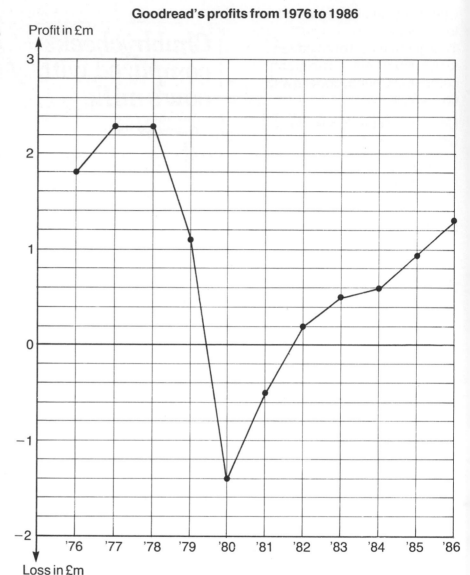

Goodread's profits from 1976 to 1986

Profit in £m

Loss in £m

Slobbertales is a rival publishing company. The table below shows Slobbertales' profits for the same ten-year period.

Slobbertales' profits from 1976 to 1986

1976	1977	1978	1979	1980	1981	1982	1983	1984	1985	1986
−1.2m	−0.4m	+0.6m	+0.9m	+1.2m	+2.7m	+2.3m	+2.2m	+2.0m	+1.7m	+1.3m

8 On the same grid as your first graph, draw another graph to show Slobbertales' profits. Draw this graph in a different colour to the first one.

9 James Gordon had offers of jobs from both of these companies in 1986. He was not sure which job to take. What advice would you have given him to help him make up his mind? Explain why you would give him this advice.

Fuel consumption

There are fifteen different models of the Vauxhall Cavalier. The graph below is taken from the Vauxhall catalogue, and shows the fuel consumption for six of the models.

For each model there are three bars indicating three different types of driving, as indicated by the key. Each bar tells you how many litres of fuel are needed to drive 100km.

Example

What fuel consumption are you likely to get from a 1300 model with a 4-speed gearbox in urban driving conditions?

The first three bars on the graph show the fuel consumption for a 1300 model with a 4-speed gearbox.
The key below the graph tells you that bar 1 is for urban driving.
The height of bar 1 is 9.1, and so the fuel consumption is 9.1 litres per 100km.

What fuel consumption are you likely to get from each car in questions 1–3?

1 An automatic 1600 saloon travelling at a steady 56 mph?
2 A 5-speed 1300 saloon travelling at a steady 75 mph?
3 An automatic 1300 saloon driving around town?
4 Would you want the bar for your model to be as low as possible or as high as possible? Explain your answer.
5 If most of the driving is around town, which of the cars in the graph would give the best fuel consumption?
6 Which model would give the best fuel consumption for motorway driving?
7 What difference does engine-size make to fuel consumption?
8 What difference does speed make to fuel consumption?
9 What is the drawback of having a 5-speed gearbox rather than a 4-speed gearbox? What are the advantages?

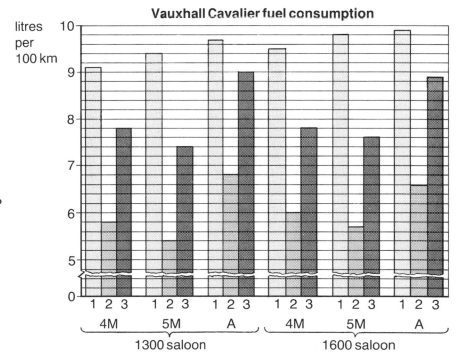

Vauxhall Cavalier fuel consumption

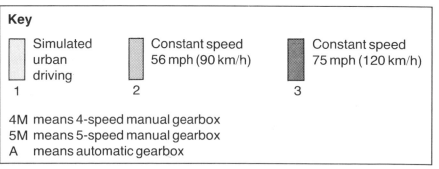

Key

- Simulated urban driving — 1
- Constant speed 56 mph (90 km/h) — 2
- Constant speed 75 mph (120 km/h) — 3

4M means 4-speed manual gearbox
5M means 5-speed manual gearbox
A means automatic gearbox

Continue the pattern

You need 0.5 cm squared paper.
Copy each of the patterns below
and continue it across your paper.
Make it about 20 cm long.

1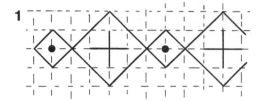

2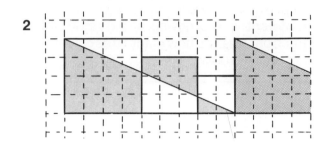

3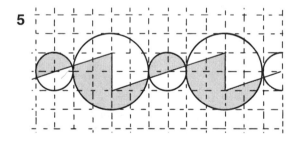

4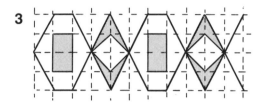

5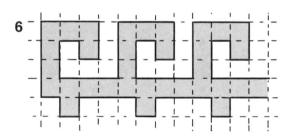

6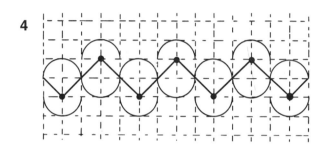

7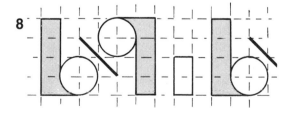

8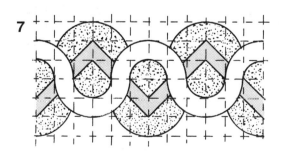

9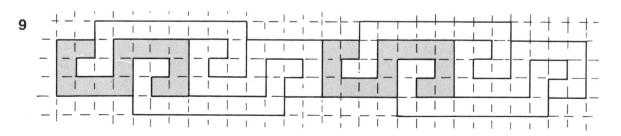

Take your chance (2)

Example

What are your chances of scoring more than 4 in one throw of a die?

Number of scores more than 4 = 2

Number of possible scores   = 6

Chances of scoring more than 4 are 2 chances in 6

(which is the same as 1 chance in 3)

Probability of scoring more than 4 is $\frac{2}{6}$

(which is the same as $\frac{1}{3}$)

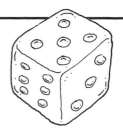

| 1 | ÷ | 3 | = | *0.333333* |

$\approx$ 0.33 (rounded correct to 2 decimal places)

Copy this table.
Make space for 8 more questions.
Fill in the table as you find your answers.

| | Chances | Probability | |
		Fraction	Decimal
Example Scoring more than 4	1 chance in 3	$\frac{1}{3}$	0.33

Round all decimals correct to 2 decimal places

1 What are your chances of scoring an even number in one throw of a die?

2 What is the probability of drawing an ace from a full pack of playing cards?

3 What is the probability of drawing any club from a full pack of cards?

4 (a) What are your chances of drawing a spotted ball from the bag on your first try?
  (b) What are your chances of drawing a striped ball from the bag on your first try?

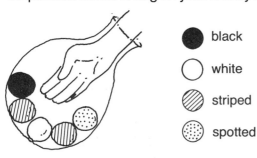

- ● black
- ○ white
- ◐ striped
- ⦿ spotted

5 You pile up these cards, shuffle them, and choose one card with your eyes shut.

50 23 45 36 81

95 90 123 113 150

(a) What are your chances of choosing a number which divides by 3?
(b) What are your chances of choosing a number which divides by 5?
(c) What are your chances of choosing a number which divides by both 3 and 5?

6 Look at the decimal numbers in the last column of your table.
Write out the eight events from the first column of your table in order of probability, from least likely to most likely.

116

More bits and pieces

Examples

This shape is divided into three equal parts.
Two of the parts are shaded.
Two-thirds of the shape is shaded.

  $\frac{2}{3}$

This shape is divided into four equal parts.
Three of the parts are shaded.
Three-quarters of the shape is shaded.

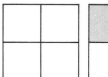 $\frac{3}{4}$

This shape is divided into six equal parts.
Five of the parts are shaded.
Five-sixths of the shape is shaded.

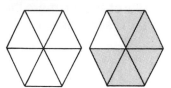 $\frac{5}{6}$

What fraction of each of the shapes is shaded?

Write the answers in figures and also in words.

1
(a) (b)

2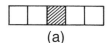
(a) (b)

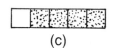
(c)

3
(a) (b) (c)

4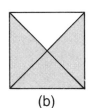
(a) (b)

5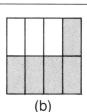
(a) (b)

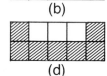
(c) (d)

6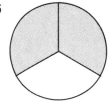

7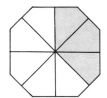

8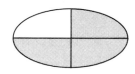

9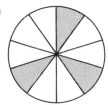

10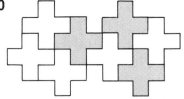

11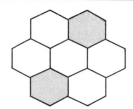

12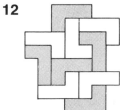

13 What fraction of this shape is covered with each shading:

(a)   (b)

(c) ▇ (d) ▨

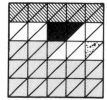

117

Copy each shape on squared paper.
Shade in the required fraction.

14 Make two copies of this shape. Shade $\frac{3}{4}$ of the first and $\frac{2}{3}$ of the second.

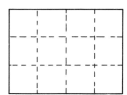

15 Two copies.
$\frac{2}{5}$ and $\frac{2}{3}$.

16 Four copies.
$\frac{5}{8}$, $\frac{7}{8}$, $\frac{3}{4}$ and $\frac{9}{16}$.

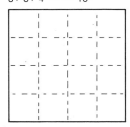

17 Three copies.
$\frac{3}{10}$, $\frac{3}{5}$ and $\frac{7}{10}$.

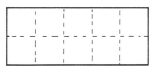

18 Two copies.
$\frac{3}{4}$ and $\frac{5}{8}$.

19 Three copies.
$\frac{5}{6}$, $\frac{2}{3}$ and $\frac{4}{6}$.

Examples

What is $\frac{3}{4}$ of £88?

£88

$\frac{1}{4}$ of £88 = £22

$\frac{3}{4}$ of £88 = 3 × £22

= £66

22 22 22 22
22 22 22 22
66

Work out $\frac{2}{3}$ of 90 kg.

90 kg

$\frac{1}{3}$ of 90 kg = 30 kg

$\frac{2}{3}$ of 90 kg = 2 × 30 kg

= 60 kg

30 30 30
30 30 30
60

20 Work out each of these:
(a) $\frac{3}{4}$ of £96
(b) $\frac{3}{4}$ of 120 people
(c) $\frac{2}{3}$ of £60
(d) $\frac{2}{3}$ of 24 hours
(e) $\frac{3}{5}$ of 60 kg
(f) $\frac{2}{5}$ of 30 litres
(g) $\frac{3}{8}$ of 1 day
(h) $\frac{5}{8}$ of £400

21 The petrol tank in this car holds 60 litres when it is full. How many litres of petrol are in the tank now?

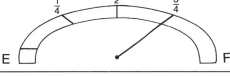

22 (a) How many small cubes are there in the large one shown here?
(b) $\frac{1}{4}$ of the cubes are painted black. How many black cubes are there?
(c) $\frac{3}{8}$ of the cubes are shaded. How many cubes are shaded?
(d) $\frac{5}{16}$ of the cubes are painted white. How many white cubes are there?
(e) The rest are shaded. How many of the cubes are shaded?

118

Starting work

Geoff leaves school and gets a job with
Mr Woods, the joiner.
Mr Woods still measures in inches.
He explains three different scales
to Geoff. ▼

$1\frac{3}{4}''$ means 'one
and three quarters
inches'.

Scale 1
Each inch is divided into 4 parts.

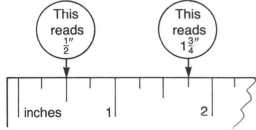

This reads $\frac{1}{2}''$

This reads $1\frac{3}{4}''$

inches   1   2

Scale 2
Each inch is divided into 8 parts.

This reads $\frac{5}{8}''$

This reads $1\frac{1}{4}''$

This reads $2\frac{3}{8}''$

inches   1   2

Scale 3
Each inch is divided into 16 parts.

This reads $4\frac{3}{16}''$

This reads $4\frac{7}{8}''$

This reads $5\frac{3}{4}''$

This reads $6\frac{9}{16}''$

4   5   6

---

1 Write down the reading at each of the points A to U on these scales.

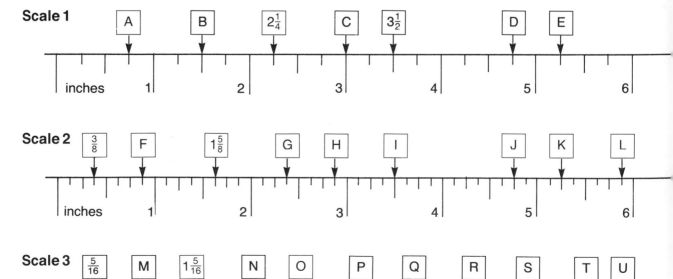

Scale 1   A   B $2\frac{1}{4}$ C $3\frac{1}{2}$ D   E

inches   1   2   3   4   5   6

Scale 2 $\frac{3}{8}$ F $1\frac{5}{8}$ G   H   I   J   K   L

inches   1   2   3   4   5   6

Scale 3 $\frac{5}{16}$ M $1\frac{5}{16}$ N   O   P   Q   R   S   T   U

inches   1   2   3   4   5   6

2 Use the 3 scales to put each set of measurements in order (from smallest to largest).

(a) $3\frac{1}{2}''$, $3\frac{5}{8}''$, $3\frac{7}{16}''$

(b) $2\frac{3}{4}''$, $2\frac{13}{16}''$, $2\frac{7}{8}''$

(c) $5\frac{3}{16}''$, $5\frac{1}{4}''$, $5\frac{3}{8}''$

(d) $1\frac{1}{2}''$, $1\frac{5}{8}''$, $1\frac{8}{16}''$

(e) $4\frac{6}{8}''$, $4\frac{3}{4}''$, $4\frac{11}{16}''$

(f) $2\frac{1}{4}''$, $2\frac{2}{8}''$, $2\frac{4}{16}''$

Your teacher will give you one of each of the 3 scales.

3 Use the scales to measure the width and thickness of each piece of wood.

(a)
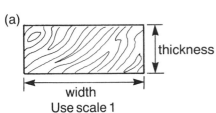
width
thickness
Use scale 1

(b)
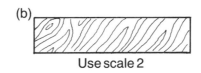
Use scale 2

(c)
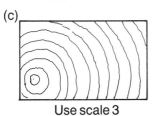
Use scale 3

(d)
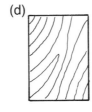

4 One of the skills Geoff had to learn was how to join pieces of wood together.
  (a) Use one of your scales to measure dimensions *a*, *b* and *c* for each of the 6 pieces in the diagram below. (*a*, *b* and *c* are only marked on piece 1.)
  (b) Pair off the pieces to show which lettered part would fit into each shaded numbered part.

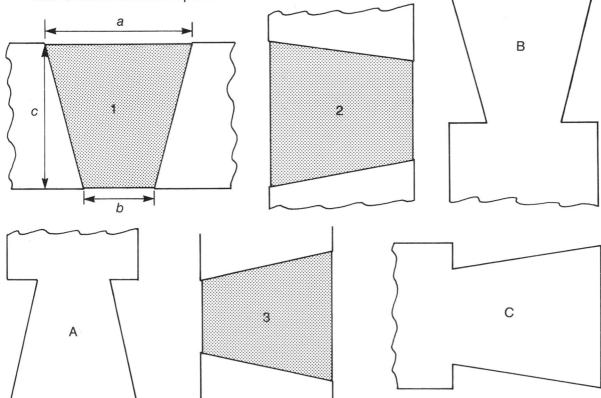

What size screw?

Sometimes Geoff has to join 2 pieces of wood together using screws.

The table opposite shows the different lengths of screws stocked in the workshop. ▶

Screws available

$\frac{3}{8}''$	$\frac{1}{2}''$	$\frac{5}{8}''$	$\frac{3}{4}''$	$1''$	$1\frac{1}{4}''$
$1\frac{1}{2}''$	$1\frac{3}{4}''$	$2''$	$2\frac{1}{2}''$	$3''$	$4''$

Example 1
What is the longest screw Geoff could use to join these 2 pieces of wood together?
(The screw must not go all the way through both pieces of wood.)

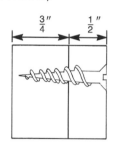

Put 2 scales together like this:

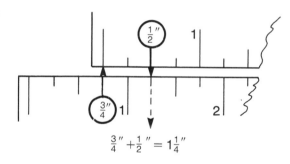

$$\frac{3}{4}'' + \frac{1}{2}'' = 1\frac{1}{4}''$$

Longest screw is $1''$.

Example 2
What screw should Geoff use to join wood $1\frac{3}{8}''$ thick to wood $\frac{7}{16}''$ thick?

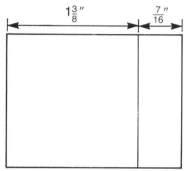

Put 2 scales together like this:

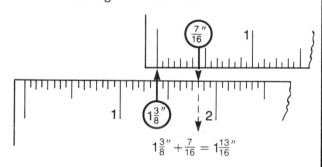

$$1\frac{3}{8}'' + \frac{7}{16} = 1\frac{13}{16}''$$

Longest screw is $1\frac{3}{4}''$.

Your teacher will give you 3 scales to help you complete this page.

T 1 Using your scales, copy and complete this table.

Thickness of 1st piece	Thickness of 2nd piece	Total thickness	Longest screw
$\frac{3}{4}''$	$\frac{1}{2}''$	$1\frac{1}{4}''$	$1''$
$1\frac{3}{8}''$	$\frac{7}{16}''$	$1\frac{13}{16}''$	$1\frac{3}{4}''$
$\frac{3}{4}''$	$\frac{3}{4}''$		
$\frac{5}{8}''$	$\frac{3}{8}''$		
$2\frac{1}{2}''$	$\frac{3}{4}''$		

Example

Geoff had to learn to make joints like this:

The 2 pieces fit together like this:

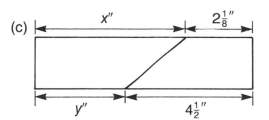

Work out the dimension *a*.

Dimension *a* is $5\frac{3}{4}'' - 1\frac{7}{8}''$

Put 2 scales together like this:

$5\frac{3}{4}'' - 1\frac{7}{8}'' = 3\frac{7}{8}''$

Dimension *a* is $3\frac{7}{8}''$

---

2 Use your scales to work out each dimension.

(a)

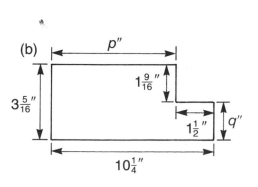

(c)

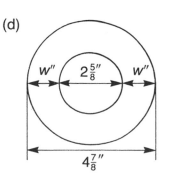

This piece of wood is $8\frac{3}{4}''$ long.

(b)

(d)

A good mix

Andy is making some mortar to build a brick wall.
The mix is 1 part cement to 3 parts sand.

Andy mixes 1 shovel of cement with 3 shovels
of sand.
He mixes all this together then pours in water.
This makes mortar for bricklaying.
Andy has mixed cement and sand in the **ratio**
1 to 3. The ratio of cement: sand is 1:3.

Example
Andy puts in 4 shovels of cement.
How many shovels of sand should he mix
with this?

Cement : sand $= 1:3$
$= 4 : 4 \times 3$
$= 4 : 12$

He should mix in 12 shovels of sand.

---

1 How many shovels of sand should Andy mix
with each of these?
   (a)   3 shovels of cement
   (b)   5 shovels of cement
   (c)   7 shovels of cement
   (d) 10 shovels of cement
   (e) 12 shovels of cement
   (f) 15 shovels of cement

2 How many shovels of cement should Andy mix
with each of these?
   (a)   6 shovels of sand
   (b) 12 shovels of sand
   (c) 15 shovels of sand
   (d) 18 shovels of sand
   (e) 21 shovels of sand
   (f) 24 shovels of sand

3 The table shows different mixes for various
types of bricklaying.

Type of wall	Cement : sand
Ordinary bricks	1 : 5
Bricks (outside wall)	1 : 3
Lightweight blocks (interior wall)	1 : 6

What type of wall is Andy building if he mixes:
(a) 5 shovels of cement with 30 shovels of
    sand?
(b) 8 shovels of cement with 24 shovels of
    sand?
(c) 5 shovels of cement with 25 shovels of
    sand?

4 To make a concrete base for a garden hut the mix is

cement : sand : gravel = 1 : 2 : 3

Andy mixes 1 bucket of cement with 2 buckets of sand and 3 buckets of gravel.
How many buckets of sand and gravel would Andy mix with
(a) 2 buckets of cement
(b) 3 buckets of cement?

5 This table tells you that one bucket of cement will make 6 buckets of concrete. Copy and complete it up to 10 buckets of cement.

Buckets of cement	1	2	
Buckets of sand	2		
Buckets of gravel	3		
Buckets of concrete	6		

Example

Andy wants to lay a concrete base for a small garden shed.
The concrete is 2 m by 1.5 m by 8 cm thick.
1 m^3 of concrete is about 50 buckets.
How many buckets of cement, sand and gravel must he mix?

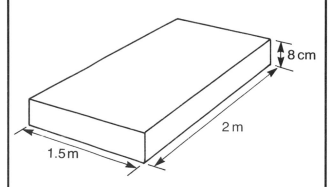

Copy this example and fill in the blanks.
Volume of concrete is
length × width × depth
= 2 m × 1.5 m × 8 cm
= 2 m × 1.5 m × 0.08 m
= . . . . m^3

Number of buckets of concrete is
. . . . × 50
= . . . .
Find this number in your table.

Andy must mix . . . . buckets of cement with . . . . buckets of sand and . . . . buckets of gravel.

6 How many buckets of cement, sand and gravel must be mixed to make each foundation? ▼

	Length	Width	Depth
(a)	1.8 m	2.5 m	8 cm
(b)	2 m	3 m	10 cm

7 Andy wants to cover an area 6 m by 4 m with slabs 50 cm square by 5 cm thick.
(a) How many slabs does he have to make?
(b) How many m^3 of concrete must he mix?
(c) How many buckets of cement, sand and gravel does he need?

No change

On many one-man-operated buses you need to have the exact fare. This saves time when there is a long queue for the bus.

If your fare is 80p and you only have a £1 coin, you must hand over the £1 coin . . . or walk!

Each of these problems is about someone wanting to travel on a one-man-operated bus.

1 Joe's fare is 35p.
The coins in his pocket are:

(a) Which coins should he give the bus driver?
(b) Joe bought a record costing £1.05.
How much money did he give the assistant?
How much change did he get?
(c) He wants to be able to pay his exact bus fare home with his change.
Which coins in his change would suit him?
Which coins in his change would *not* suit him?

2 Mrs White has these coins in her purse:

(a) If her fare is 85p, how much does she have to pay?
(b) Write down your comments about this.
(c) Just before she paid over the money a friend gave her change for one of the coins she had and she was able to pay the correct fare.
What change did her friend give her?

3

Two twenty-eights and a fifteen, please.

Mr and Mrs Robinson must pay 28p each, and a further 15p for their son. What coins must Mr Robinson give the bus driver? He has a £5 note and these coins:

4 Sue was taking her little sister into town to buy a birthday present. This is the money she had:
Sue must pay 55 pence for herself and 27 pence for her sister.
(a) How much must Sue give the bus driver?
(b) Sue bought her little sister this doll.
Will Sue have the correct money for the fares home? Explain how you know.

£4.45

5 Mrs Adams and Mrs Brownlee are neighbours and very good friends. They meet at the bus stop and discover they are going on the same journey. The fare is 67p each.
They have the coins shown opposite:

(a) Explain why Mrs Adams must pay extra if she pays her own fare.

(b) How can they work things out between them so that neither has to pay the driver extra?
Find two ways of doing this.

(c) There is one way that each lady can pay the driver 67p and get the correct change back before leaving the bus. How can this be done?

Mrs Adams

Mrs Brownlee

Mrs Adams' coins

Mrs Brownlee's coins

Cheque change

This notice appears at the entrance to the Fab Fare supermarket.

FAB FARE

Instructions to customers wishing to pay by cheque:
1 round your bill up to the next £
2 make out a cheque for that amount
3 the cashier will give you your change.

£50 limit

Example
Mrs Farmer's bill came to £46.37. Here is her cheque:
Mrs Farmer's change was 63p.

Following the instructions, write out cheques for each of these bills. What change should each customer expect?

1 Fiona Anderson's bill came to £39.52.

2 Jason Ansari's bill was £13.39.

3 Miss Goldstein's bill was £30.28.

4 Mr Bradford's bill came to £7.89.

5 Parminder Bedi's bill came to £78.45.

6 Peter Wilson's bill came to £15.16.

7 Laurence Wighton's bill came to £25.46.

8 Nadia Darcy's bill came to £19.99.

9 Jarmin Kahn's bill came to £68.47.

Any change?

Example
Bina bought these items at the supermarket.
She has only £6 in her purse. Does she have
enough money?

Use the flowchart below.

Find the total.	£5.52
Does she have enough money?	yes
Work out the change.	£6.00
	−£5.52
	£0.48

Bina has enough money and gets 48p change.

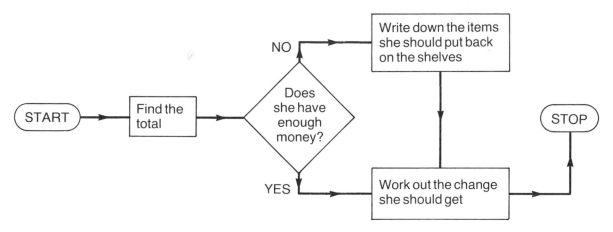

Follow the flowchart for each of these:

1 Mr Appleton has £10. This is his bill at the checkout.

```
1.56
0.73
0.19
3.99
0.38
0.83
1.17
        TOTAL
```

2 Sophie has £20. This is her bill at the checkout.

```
3.99
0.57
0.84
1.49
0.39
3.50
7.99
1.57
        TOTAL
```

3 Ali is buying some items for his shop. He has only £10 with him. He has left his cheque book at home.

4 Dave's mum sent him to the supermarket with £6 to buy five 100W pearl light bulbs, three 60W clear, two 150W clear, and two red bulbs for the gas fire.

	Light bulbs		
	Clear	**Pearl**	**Coloured**
60 watts	35p	38p	75p
100 watts	40p	44p	–
150 watts	45p	50p	–

Short of change

I'm sorry, sir, we're very short of change. Do you have 10p? I will give you £2 change.

Explain to your teacher how this works.

What might the person at the checkout say to you for each of these items when she is short of change?

1

2

3 What would the checkout girl say in each of these cases?

		Cash offered			**Cash offered**
(a)	£3.09	£ 5	(d)	£11.05	£15
(b)	£2.11	£ 3	(e)	£12.28	£20
(c)	£7.15	£10	(f)	£15.09	£20

Sometimes this may happen at the checkout.

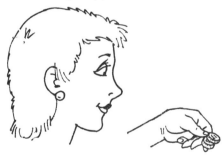

I'm very sorry. I have nothing smaller than 50p. Do you have 15p? I will give you 50p change.

Explain to your teacher how this works.

Work out the total and correct change for each bill.
What might the person at the checkout say to you for each of these bills when she has nothing smaller than 50p?

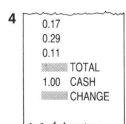

4
```
 0.17
 0.29
 0.11
▓▓▓ TOTAL
 1.00 CASH
▓▓▓ CHANGE
```

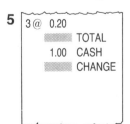

5
```
3@  0.20
   ▓▓▓ TOTAL
 1.00  CASH
   ▓▓▓ CHANGE
```

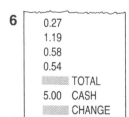

6
```
 0.27
 1.19
 0.58
 0.54
▓▓▓ TOTAL
 5.00  CASH
▓▓▓ CHANGE
```

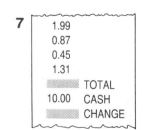

7
```
 1.99
 0.87
 0.45
 1.31
▓▓▓ TOTAL
10.00  CASH
▓▓▓ CHANGE
```

Keep the change!

Sue works every Saturday from 10 am until 4 pm in Tina's Tea-room. Her job is to serve the customers, deal with their bills and clear the tables.

One Saturday she was in luck! *Every* customer said, 'Keep the change'.

The customers Sue served between 12.30 and 1 pm are shown below. Work out
(a) each customer's bill
(b) Sue's tip from each customer.

TINA'S TEA ROOM
PRICE LIST

Coffee	33p	Milk	20p
Tea	30p	Cola	25p
Cakes	26p	Fresh Orange	30p
Chocolate Biscuit	12p	Chips	35p
Doughnut	28p		

1 Mrs Dale

A cup of tea, a cake, and a chocolate biscuit, please dear.

2 Mr and Mrs Murphy

Two portions of chips, one cola, one fresh orange, and two doughnuts, please.

3 Clive, Penny and Hassan.

Three glasses of fresh orange, two cakes, and one doughnut, please.

4 A party of 10 tourists from Aberdeen.

```
⊖ TINA'S TEA ROOM ⊖
3 teas
3 coffees
2 glasses milk
2 colas
1 orange
5 cakes
5 biscuits
          _____
TINA'S TOTAL
```

5 Mr and Mrs Samad and their 2 children.

```
⊖ TINA'S TEA ROOM ⊖
2 teas
2 coffees
4 portions chips
1 cake
3 doughnuts
          _____
TINA'S TOTAL
```

6 Five workmen.

```
⊖ TINA'S TEA ROOM ⊖
5 teas
5 portions chips
5 doughnuts
5 chocolate biscuits

TINA'S TOTAL
```

7 How much altogether did Sue make in tips between 12.30 and 1 pm?

8 How much do you think Sue would make in tips between 10 am and 4 pm? Explain your answer.

9 Sue saves her tips to buy a portable TV costing £89.99. How many weeks will it take her to save enough money?

Quick change

1 Copy this grid.
Change CASH into BILL using these two rules:
  (i) Change only one letter at a time.
  (ii) Each new line must be a proper English word.

The shapes on the right will help you. Copy the shapes on to a sheet of paper. Cut out the shapes and fit them together.
The words CASH and BILL are included on the shapes.

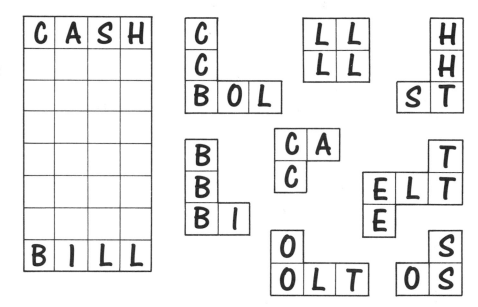

2 This is a child's toy called 'Change the face'.
The game pieces include two heads, two pairs of eyes, two pairs of ears, two mouths, and two noses.

How many different faces can you make from these pieces? The eyes must match and the ears must match.

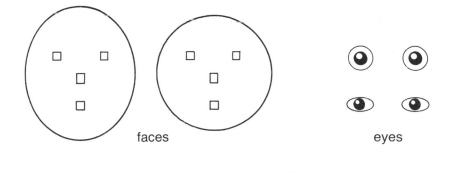

faces
eyes
mouths
noses
ears

Here are some to start you off:

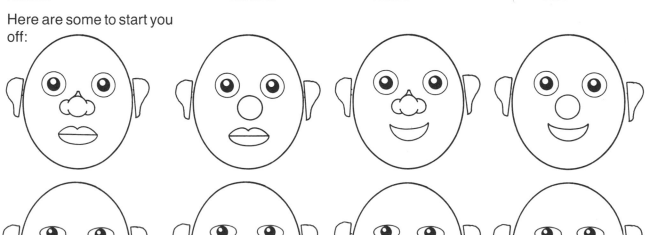

Time words

Answer these questions about time.

1 (a) What is a time-traveller?
(b) Which film does this photo come from?

2 How do you *kill* time?

3 Where would you *serve* time?

4 What is a pastime?

5 What is a time-bomb?

6 What is a two-timer?

7 How do you *take* time *off*?

8 What do these words mean?
(a) overtime   (b) double time   (c) part-time

9 How do you *make* time?

10 Who *beats* time?

11 Where would you see this?

12 What does Joe's mum mean?

It's high time you looked for a job!

13 What is half-time?

14 What is time-out?

15 What is a time-zone?

16 What does it mean if you are a good time-keeper?

17 What does time-sharing mean?

18 What is a time-switch?

How long?

Find the answers to these time questions.

1 How long does it take to boil an egg?

2 You are a world class athlete. How long does it take you to run 100 metres?

3 By what age should a child normally be able to walk?

4 Karen's mother takes a frozen chicken out of the fridge.
For how long should she leave the chicken to defrost before cooking it?

5 For how long are the players on the field in a normal football game?
For how long are they off the field at half-time?

6 For how long does a coin spin on the top of your desk before it falls flat?

7 Which of these piecharts shows how long you are asleep each day? (Each circle represents a whole day.)

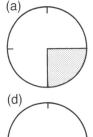
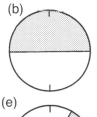
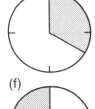
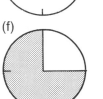

(a) (b) (c)

(d) (e) (f)

 asleep

 awake

8 For how long is a human baby inside its mother before birth?

9 How long ago did the second world war begin? How long ago did it end?

10 How many hours do you spend watching TV each week?

11 (a) How long does the Earth take to travel once round the Sun?
(b) How long does the Moon take to travel once round the Earth?

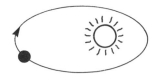

12 How long does it take this pendulum to make one complete swing away and back. (Time 10 swings, then divide by 10.)

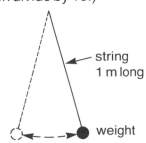

string 1 m long

weight

13 For how long do you go without blinking your eyelids? (Ask a friend to time you when you are not thinking about it.)

A busy woman

Ms Rice is a director of a large engineering firm.
This is a page from her appointments book.
The pictures below show six scenes from Ms Rice's day on Wednesday.

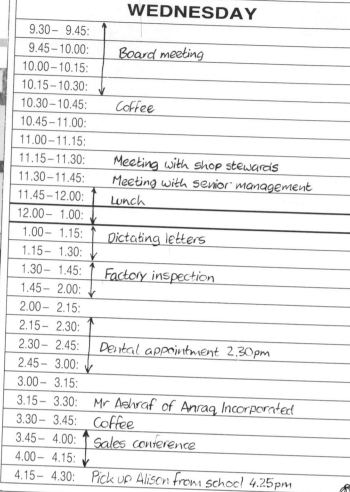

WEDNESDAY	
9.30 – 9.45:	
9.45 – 10.00:	Board meeting
10.00 – 10.15:	
10.15 – 10.30:	
10.30 – 10.45:	Coffee
10.45 – 11.00:	
11.00 – 11.15:	
11.15 – 11.30:	Meeting with shop stewards
11.30 – 11.45:	Meeting with senior management
11.45 – 12.00:	Lunch
12.00 – 1.00:	
1.00 – 1.15:	Dictating letters
1.15 – 1.30:	
1.30 – 1.45:	Factory inspection
1.45 – 2.00:	
2.00 – 2.15:	
2.15 – 2.30:	
2.30 – 2.45:	Dental appointment 2.30pm
2.45 – 3.00:	
3.00 – 3.15:	
3.15 – 3.30:	Mr Ashraf of Anraq Incorporated
3.30 – 3.45:	Coffee
3.45 – 4.00:	Sales conference
4.00 – 4.15:	
4.15 – 4.30:	Pick up Alison from school 4.25pm

1 Put the correct pictures and times together.

(a)

(b)

(c)

(d)

(e)

(i)
(ii)
(iii)
(iv)
(v)

2 Mr Jones answers all Ms Rice's telephone calls.
What would you say if you were Mr Jones?
Each caller wanted to speak to Ms Rice.

(a)

(b)

(c) (d) (e)

3 Write down an accurate time for each of these phone calls.

(a) I'm sorry, she went into a sales conference five minutes ago, so she won't be available for some time.

(b) She has just left to pick up her daughter from school.

(c) If you hang on for two minutes she will be out of the board meeting.

(d) She is ten minutes late coming back from the dentist. Could you phone again later, please?

(e) She has just come out of a meeting, but she has another one immediately. Hold on and I will try to contact her.

In all these questions assume that Ms Rice kept to her schedule for all appointments.

4 Ms Rice's working day is from 9.30 am to 4.30 pm. How long is this?

5 How long did she spend at lunch and coffee breaks?

---

6 How much time did Ms Rice spend at meetings on Wednesday? (Count the factory inspection as a meeting.)

7 How much time did she spend in the office and factory?

8 Ms Rice usually does paper work when she has no appointment. How long did she spend on paperwork on Wednesday? (Dictating letters counts as paperwork.)

---

9 Trace this circle. Draw a piechart to show how Ms Rice spent Wednesday. The labels on your piechart should be:
Meetings
Lunch
Personal
Paperwork.
Each section represents 15 minutes of Ms Rice's day.

10 What fraction of the day did Ms Rice spend on lunch and coffee breaks?

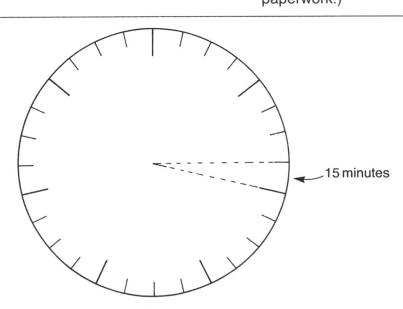

15 minutes

Flight times

These clocks are in the lounge of an airport hotel.

 London GMT — Sept 15   pm

Paris — Sept 15   pm

 Hong Kong — Sept 16   am

Los Angeles — Sept 15   am

 New York — Sept 15   pm

T Your teacher will give you a page of blank clock faces. For each of questions 1 to 4 you will need five clock faces (one for each of the five cities above). In each question, copy the given clock face, and then fill in the time and date for the other four cities.

1 London GMT — Dec 20   pm

2 Hong Kong — May 1   am

3 New York — Oct 17   pm

4 Los Angeles — Jan 1   am

5 Mr Wiseman is due to fly from London to New York on business. His watch, opposite, shows the time and date of his flight. He arrives in New York 7 hours later.

   (a) Draw sketches to show the time and date on his watch when he arrives in New York, before he alters his watch.

   (b) He must now change his watch to show New York time. Draw more sketches to show the new time and date.

time

date (Nov 2)

Repeat question 5 for each of these flights.

	Point of departure	Departure time	Date	Flight time	Destination
6	New York	8.00 am	Nov 7	7 hours	London
7	London	11.00 am	Apr 3	16 hours	Hong Kong
8	Hong Kong	3.00 am	Apr 9	16 hours	London
9	Paris	9.00 am	Mar 10	12 hours	Los Angeles

10 Mr Wiseman caught a 2.00 pm flight from Heathrow to New York on 17 October. The airport clock on the right shows the time when he arrived in New York.

He leaves New York for Los Angeles after a break of $1\frac{1}{2}$ hours. The second airport clock shows the time when he arrived in Los Angeles. Work out:

   (a) the flight time from London to New York
   (b) the flight time from New York to Los Angeles
   (c) Mr Wiseman's total flying time.

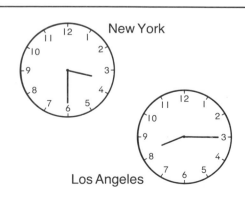
New York

Los Angeles

Beach blunder

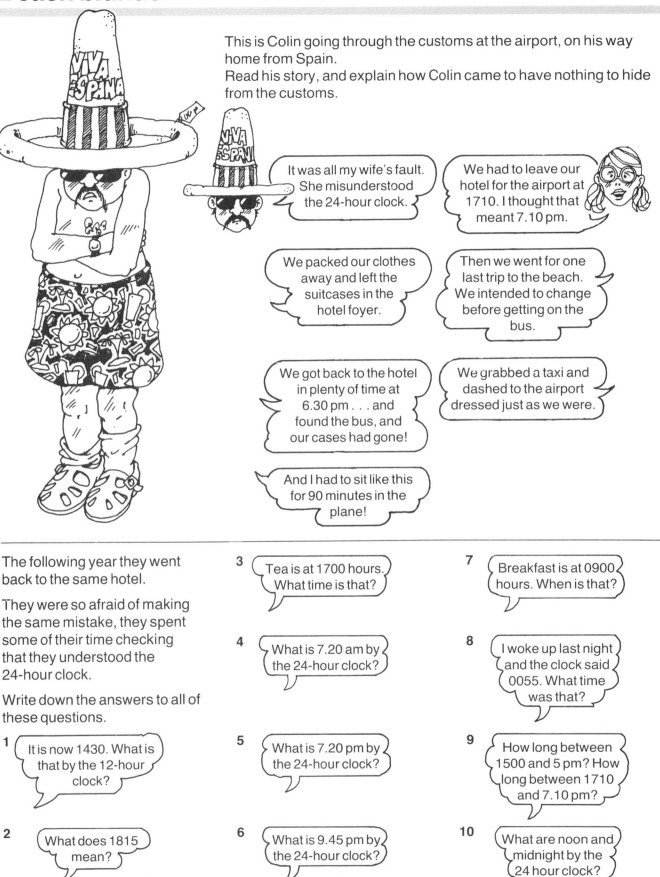

This is Colin going through the customs at the airport, on his way home from Spain.
Read his story, and explain how Colin came to have nothing to hide from the customs.

It was all my wife's fault. She misunderstood the 24-hour clock.

We had to leave our hotel for the airport at 1710. I thought that meant 7.10 pm.

We packed our clothes away and left the suitcases in the hotel foyer.

Then we went for one last trip to the beach. We intended to change before getting on the bus.

We got back to the hotel in plenty of time at 6.30 pm . . . and found the bus, and our cases had gone!

We grabbed a taxi and dashed to the airport dressed just as we were.

And I had to sit like this for 90 minutes in the plane!

The following year they went back to the same hotel.

They were so afraid of making the same mistake, they spent some of their time checking that they understood the 24-hour clock.

Write down the answers to all of these questions.

1 It is now 1430. What is that by the 12-hour clock?

2 What does 1815 mean?

3 Tea is at 1700 hours. What time is that?

4 What is 7.20 am by the 24-hour clock?

5 What is 7.20 pm by the 24-hour clock?

6 What is 9.45 pm by the 24-hour clock?

7 Breakfast is at 0900 hours. When is that?

8 I woke up last night and the clock said 0055. What time was that?

9 How long between 1500 and 5 pm? How long between 1710 and 7.10 pm?

10 What are noon and midnight by the 24 hour clock?

Hot air

This time-clock controls the central heating.
Indicators 1 and 3 switch the heating *on*.
Indicators 2 and 4 switch the heating *off*.
The dial turns in the direction of the large arrow.

Here are 7 things we can tell from this clock:
  (i) The heating is on from 6.15 am until 9.30 am.
  (ii) The heating is off from 9.30 am until 4.30 pm.
  (iii) The heating is on from 4.30 pm until 9 pm.
  (iv) The heating is off from 9 pm until 6.15 am.
  (v) The heating is on for 7 hours 45 minutes
      each day.
  (vi) The time is 2.15 pm.
  (vii) The heating will come on again in 2 hours 15
      minutes.
Check with your teacher that you understand
these 7 things.

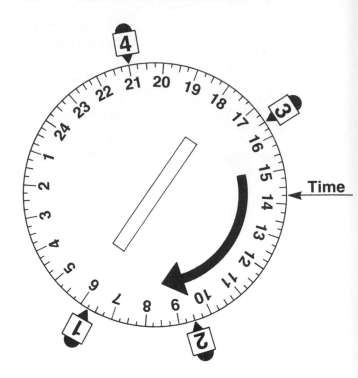

Write down 7 things you can tell from each of the
time-clocks below. Use the example above to
help you.

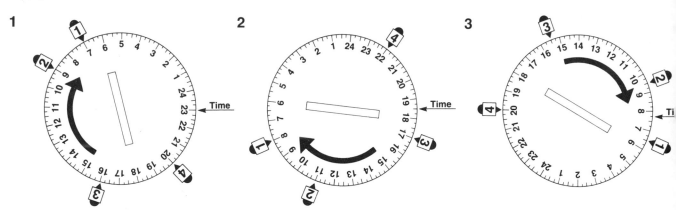

T Your teacher will give you a page of time-clocks to help you with the
following questions.
Cut out a time-clock for each question and paste it into your notebook.

4 (i) Draw in the indicators for
      each clock in the table.
   (ii) For each of the clocks,
      say how long the heating
      is on each day.

Clock	On	Off	On	Off	Time
(a)	6.00 am	9.30 am	4.00 pm	7.30 pm	3.00 am
(b)	6.30 am	9.00 am	4.30 pm	8.30 pm	7.00 pm
(c)	7.15 am	9.45 am	3.45 pm	7.45 pm	12 noon
(d)	8.45 am	10.15 am	4.30 pm	9.15 pm	8.45 am
(e)	8.15 am	10.45 am	4.15 pm	8.45 pm	11.15 pm
(f)	7.30 am	9.30 am	4.30 pm	6.30 pm	2.30 pm
(g)	7.15 am	8.45 am	4.15 pm	8.30 pm	5.15 pm

5 This clock keeps the heating switched on for 7 hours a day. Draw these three indicators and add the missing indicator on one of the time-clocks you have been given.

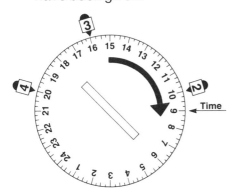

6 This clock has the heating switched on for 8 hours a day. Draw all four indicators on one of the time-clocks you have been given.

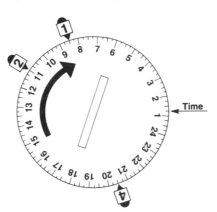

7 This clock keeps the heating switched on for six and a half hours each day. Draw in the missing indicator as before.

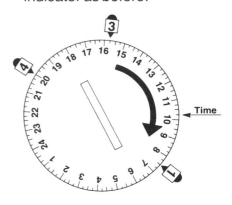

8 This clock keeps the heating on for 7 hours 30 minutes a day. Draw in the missing indicator as before.

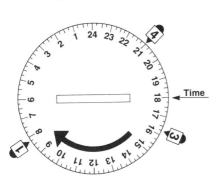

9 This clock has the heating on for 5 hours 15 minutes a day. Draw in the missing indicator as before.

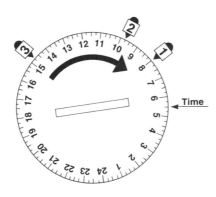

10 This clock keeps the heating on for 5 hours a day. The heating is on for the same length of time both in the morning and the afternoon. Draw in all four indicators.

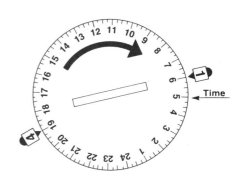

11 Cut out a clock for each set of clues.
The indicators must be drawn in the correct place and each clock set at the correct time.

Clock (a)
This clock switches on the heating at 7.30 am and 4.30 pm.
The heating goes off at 9 am and is on for 5 hours per day.
The heating came on for the morning an hour and a half ago.

Clock (b)
This clock keeps the heating on for 4 hours per day.
Indicator 1 is set at 7 am, indicator 2 at 9 am and indicator 4 at 6 pm.
The heating will go off for the evening in 1 hour.

Clock (c)
This clock keeps the heating on for 8 hours 30 minutes a day.
Indicator 1 is set at 6.30 am, indicator 2 at 8.45 am, and indicator 3 at 4.15 pm.
The heating went off 2 hours ago. It will come back on again in $5\frac{1}{2}$ hours time.

Car park

Mark's dad parks the car in a multi-storey car park every Saturday while the family goes shopping.
He does not like to pay any more than 50p.
A card with the time stamped on it is issued by a machine as the car enters.
An attendant works out the price to be paid as the car goes out.

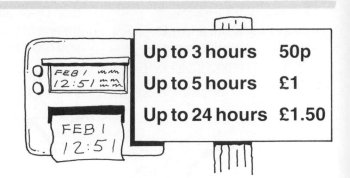

Up to 3 hours	50p
Up to 5 hours	£1
Up to 24 hours	£1.50

1 Follow the flowchart for each ticket.

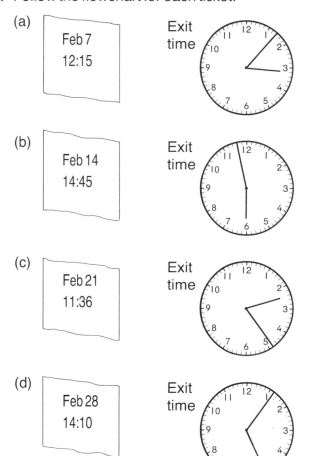

(a) Feb 7   12:15   Exit time

(b) Feb 14   14:45   Exit time

(c) Feb 21   11:36   Exit time

(d) Feb 28   14:10   Exit time

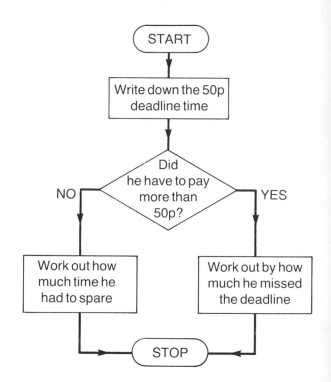

START

Write down the 50p deadline time

Did he have to pay more than 50p?

NO — Work out how much time he had to spare

YES — Work out by how much he missed the deadline

STOP

2 Copy this table, then fill it in.

Work out how much parking time Mark's dad got for his money in February.
Give your answer in minutes per penny (to one decimal place).

February			
Date	Time parked (hours/minutes)	Time parked (minutes)	Cost
7th	2 hrs 52 mins	172	£0.50
14th			
21st			
28th			
Totals			

Playing time

Copy this table, then fill in the missing numbers in
the seconds row.

Minutes	1	2	3	4	5	6	7	8	9	10
Seconds	60	120			300					

Example 1
What is the total playing time
of this tape?
(17.48 means Side 1 plays for
17 minutes 48 seconds.)

Rock on!	
Side 1	17.48
Side 2	16.39

Minutes	Seconds	
17	48	
+16	39	
33	87	$87 = \mathbf{60} + 27$
33+1	27	$= 1$ min 27 sec

Total playing time is 34 minutes 27 seconds.

Example 2
Copy and complete this example.
Find the total playing time.

Double album	
Side 1	15.45
Side 2	16.58
Side 3	16.05
Side 4	15.51

Minutes	Seconds	
15	45	
16	58	
16	05	
15	51	
?	159	$159 = \mathbf{120} + ?$
?+2	?	$= 2$ min ? sec

Total playing time is ＿＿ minutes ＿＿ seconds.

Work out the total playing time of each of these tapes.

1
Heavy Metal	
Side 1	16.35
Side 2	15.17

2
Love Songs	
Side 1	15.35
Side 2	15.36

3
Swing Time	
Side 1	17.45
Side 2	16.54

4
The Beatles	
Side 1	15.48
Side 2	16.08
Side 3	15.52
Side 4	16.27

5
Wham	
Side 1	14.57
Side 2	15.08
Side 3	16.18
Side 4	15.43

6
Jazz Hits	
Side 1	16.43
Side 2	15.39
Side 3	16.00
Side 4	15.37

7
Best of Brass	
Side 1	
Track 1	4.05
Track 2	3.30
Track 3	4.40
Track 4	4.56
Track 5	5.10
Side 2	
Track 1	4.20
Track 2	3.58
Track 3	4.10
Track 4	5.07
Track 5	4.25

8
Requiem	
Andrew Lloyd Webber	
Side 1	
1 Requiem & Kyrie	6.39
2 Dies irae	
(a) Dies irae	6.00
(b) Recordare	3.21
(c) Ingerisco	7.40
Side 2	
1 Offertorium	5.19
2 Hosanna	4.51
3 Pie Jesu	3.53
4 Lux aeterna	
& libera me	7.29

9
Oasis	
Peter Skellern	
Side 1	
Prelude	2.14
If this be the last time	4.18
I wonder why	3.51
Hold me	4.10
Oasis	5.39
Side 2	
Sirocco	6.17
Who knows	4.55
Weavers of moonbeams	5.01
Loved and lost	5.13
True love	4.33

Alarm call

Example 1
This radio alarm clock shows 24-hour time. How long until the alarm goes off?

Time	Alarm set

The time is 10.30 pm.

The radio will come on at 7.30 am and stay on for 1 hour.

Time until midnight

hours	mins		hours	mins
24	00		23	60
−22	30	→	−22	30
			1	30

	hours	mins
Time until midnight	1	30
Midnight until alarm	+7	30
Total time to alarm	9	00

The alarm will go off in 9 hours.

Example 2
How long until the alarm goes off?

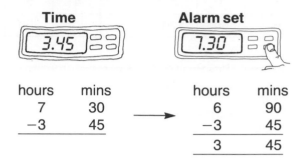

hours	mins		hours	mins
7	30		6	90
−3	45	→	−3	45
			3	45

The alarm will go off in 3 hours 45 minutes.

How long until each of these alarms goes off?

1
Time	Alarm set
5.15	6.30

2
Time	Alarm set
2.55	7.15

3
Time	Alarm set
23.40	8.00

4
Time	Alarm set
22.48	6.45

5 Sasha says she needs 10 hours sleep every night. How long before she should be asleep?

Time	Alarm set
21.35	7.45

6 Tim needs only 8 hours sleep. How long before he should be asleep?

Time	Alarm set
11.25	8.15

7 This is Danny setting the alarm on Tuesday evening but on Wednesday morning the radio alarm does not wake him!

Alarm set

(a) How long does he have before the radio switches off?

Danny wakes up 10 minutes after the radio switches off and leaves the house 12 minutes later. It takes him exactly 15 minutes to get to school which starts at 9.00 am. His teacher is 8 minutes late in arriving and marks the register 2 minutes later.

(b) Does Danny get marked late or not?

Cooling down

This is an experiment Karen did in her science lesson.

1 Heat the water in the beaker.
2 Take the temperature of the water.
3 Allow the water to cool, taking its temperature every 5 minutes.

Karen drew this graph of her results. Answer these questions using Karen's graph.

1 What was the temperature of the water at its hottest?
2 How hot was the water after 5 minutes?
3 How many degrees did it cool in the first 5 minutes?
4 What was its temperature after a quarter of an hour?
5 What was its temperature after 20 minutes?
6 Through how many degrees did it cool in the second five minutes?
7 After how many minutes had it cooled to 30°C?
8 After how many minutes had it cooled to 17°C?
9 What do you think was the temperature of the room?
10 Copy and complete the table opposite then fill it in.
11 Write a few sentences to describe how fast the water cooled at different times during the experiment.

12 This graph shows the results of another experiment Karen did using a beaker of water. Describe the experiment.

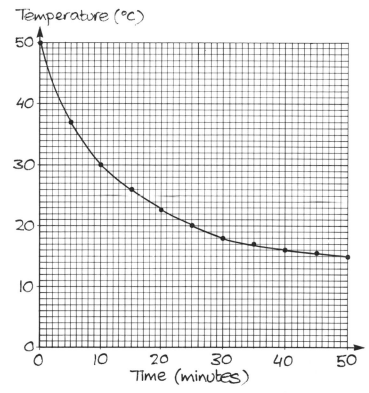

WATER COOLING   Karen

Time (minutes)	0	5	10	15	⌇	45	50
Temperature	50°	37°			⌇		
Time interval	—	0–5	5–10	10–15	⌇	40–45	45–50
Drop in temperature	—	13°			⌇		

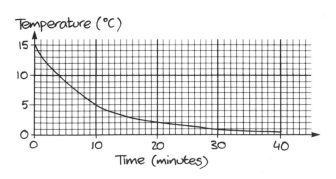

Peculiar prices

1 Write down your comments on this customer. Explain anything you say about him.

How much would you expect to pay for
(a) 10 boxes (b) 15 boxes
(c) 8 boxes (d) 12 boxes
(e) 14 boxes (f) 16 boxes

2 You are in charge of pricing plants at the nursery.
What price would you mark on this label?
Explain how you arrived at your answer.

Using your price, how much would you charge for
(a) 10 plants (b) 12 plants?

3 Nita was puzzled by this notice in the ironmonger's window.

What was the ironmonger selling?

4 What advice would you give Stephen and his mum and dad?

Mug shots

Last week the photographer was in school.
The price list which everyone got with their photographs is shown below.

Price list

The largest print may be purchased for £2.35.

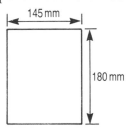
145 mm
180 mm

The smaller prints may be purchased for 90p each.

90 mm
110 mm

The wallet prints may be purchased for 50p each.

60 mm
75 mm

If you wish to buy all the photographs the total price is only £4.50.

How much should each of these parents pay?

1
I'll buy the lot Karen.

2 *You can get one of each, Josh.*

3 *I'll take everything except the large print, Sue.*

4 Alamdar's mother sent back the 3 wallet prints and paid for the rest.

5 This is what Jean's mum said to her.

'I'm sorry, Jean. I can only afford the big one, the smaller prints, and *one* of the wallet prints.'

What would you say to Jean's mum?

6 Make a good drawing to show how the photographer can make a complete set of prints on a rectangular sheet of photographic paper without wasting any paper. What size is this sheet?

144

Papering a room

This table shows how many rolls of wallpaper you need for different sizes of room.

Work out the number of rolls needed and the cost for each room in questions 1–6.

Height from skirting in metres	Distance around room in metres (including doors and windows)												
	10	11	12	13	14	15	16	17	18	19	20	21	22
2.0 – 2.2	5	5	5	6	6	7	7	7	8	8	9	9	10
2.2 – 2.4	5	5	6	6	7	7	8	8	9	9	10	10	10
2.4 – 2.6	5	6	6	7	7	8	8	9	9	10	10	11	11
2.6 – 2.8	6	6	7	7	8	8	9	9	10	11	11	12	12
2.8 – 3.0	6	7	7	8	8	9	9	10	11	11	12	12	13
3.0 – 3.2	6	7	8	8	9	10	10	11	11	12	13	13	14
3.2 – 3.4	7	7	8	9	9	10	11	11	12	13	13	14	15

Example

How much does it cost to wallpaper a room which is 4 m long, 3 m wide and 2.3 m high, if the wallpaper costs £4.50 per roll?

Height 2.3 m, 3 m, 4 m

Distance round room is
4 m + 3 m + 4 m + 3 m
= 14 m
Number of rolls is 7 (from table)
Cost is 7 × £4.50
= £31.50

1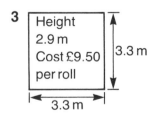
Height 2.15 m
Cost £7.50 per roll
3 m
5 m

2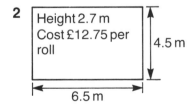
Height 2.7 m
Cost £12.75 per roll
4.5 m
6.5 m

3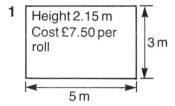
Height 2.9 m
Cost £9.50 per roll
3.3 m
3.3 m

4
Height 2.5 m
Cost £4.99 per roll
Scale 1 cm to 1 m

5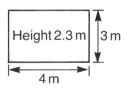
Height 2.25 m
Cost £6.30 per roll
Scale 1 cm to 2 m

6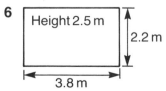
Height 2.5 m
2.2 m
3.8 m

This room cost £21 to paper. What was the cost per roll?

DIY

Dave's dad wants to replace this door with two louvre doors. The louvre doors are hinged and must meet at the centre.

The table below lists the prices of louvre doors.
If there is not a door exactly the size you need, you must buy the next biggest size and plane it down.

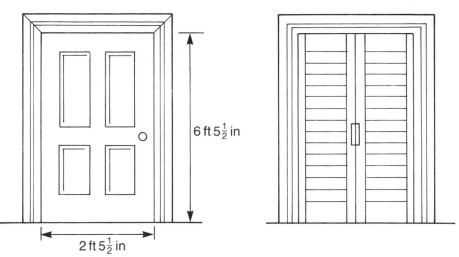

6 ft 5$\frac{1}{2}$ in

2 ft 5$\frac{1}{2}$ in

Height × width	Price	Height × width	Price	Height × width	Price
18" × 12"	£1.50	24" × 18"	£4.25	72" × 18"	£10.95
18" × 15"	£2.95	30" × 15"	£4.75	72" × 21"	£11.95
18" × 18"	£3.50	30" × 18"	£5.50	72" × 24"	£12.95
24" × 12"	£3.25	66" × 18"	£10.20	78" × 15"	£10.50
24" × 15"	£3.75	72" × 15"	£9.95	78" × 18"	£11.95

1 (a) What size doors must Dave's dad buy? (1 foot = 12 inches)
 (b) How much will they cost him?

2 Dave wants to replace two small cupboard doors in his bedroom with four louvre doors. Each cupboard is 2 ft high and 2 ft 6 in wide. Dave's dad says he will pay £10 towards the cost of the doors. How much does Dave have to pay?

3 Each of these doors is to be replaced with two louvre doors.
Work out the size of doors needed and the cost.

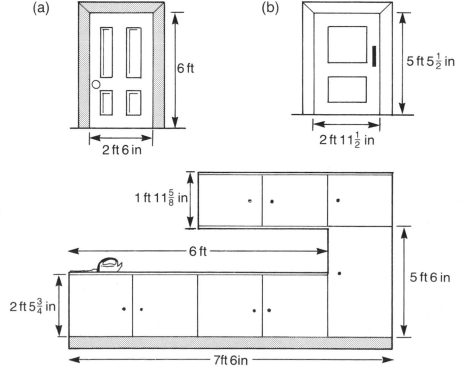

(a)

6 ft

2 ft 6 in

(b)

5 ft 5$\frac{1}{2}$ in

2 ft 11$\frac{1}{2}$ in

4 Dave's dad was so pleased with his louvre doors that he decided to put them on all his kitchen units.
 (a) What doors does he need?
 (b) What will they cost?

1 ft 11$\frac{5}{8}$ in

6 ft

5 ft 6 in

2 ft 5$\frac{3}{4}$ in

7ft 6in

Send me a letter

These postage rates were in operation at Christmas time 1986 for letters and cards:

Inland (United Kingdom, Channel Islands, Isle of Man and Irish Republic)

	First class	Second class
60 g	17p	13p
100 g	24p	18p
150 g	31p	22p
200 g	38p	28p

Airmail to Europe and surface mail to all other countries

20 g	22p
60 g	37p
100 g	53p

Airmail to countries outside Europe

	First 10 g	Each extra 10 g
Zone A	29p	11p
Zone B	31p	14p
Zone C	34p	15p

Zone A N Africa, Middle East
Zone B Americas, Africa, India, S E Asia
Zone C Australasia, Japan, China

1 How much does it cost to post each of these letters?
(a) A first class letter to Liverpool weighing 45 g.
(b) A second class letter to Poole weighing 90 g.
(c) A first class letter to Dublin weighing 175 g.
(d) A 10 g airmail letter to Washington DC.
(e) An airmail letter to Yokohama weighing 18 g.
(f) An airmail letter to Paris weighing 15 g.

2 You are going to make a chart to show all the postage rates at a glance.

You need an A4 sheet of squared paper turned this way:

The chart is started for you on the right:
Copy the chart.
Go up to 200 g for UK mail. Go up to 100 g for the others. Fill in all the prices.

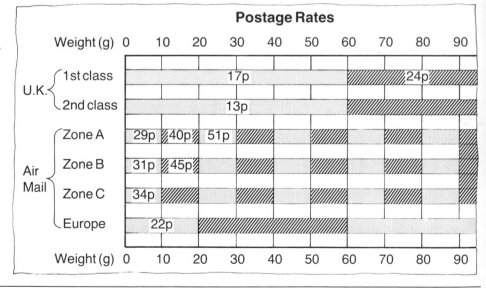

Use your chart to answer these questions.

3 What does it cost to post a 35 g airmail letter to Australia?

4 What is the heaviest first class letter you can post to somewhere in the UK for 31p?

5 A 65 g airmail letter was posted for £1.15. To which zone was it posted?

6 What does it cost to send a 175 g, second class letter to Penzance?

7 What does it cost to send a 45 g airmail letter to India?

8 A 140 g first class letter to Belfast costs the same as an airmail letter to Washington DC. What weight is the airmail letter?

9 Use your chart to find the cost of posting each letter. (If a UK letter does not say 'first class', it is second class.)

(a)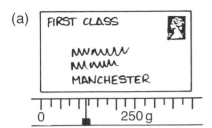

(b)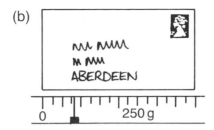

(c)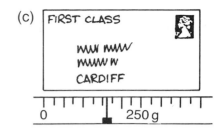

(d)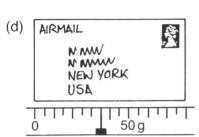

(e)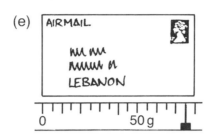

(f)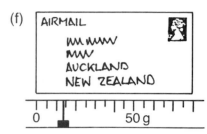

The price is right

In a TV game show contestants have to choose goods from the stall to the value of not more than £20.

The person closest to £20 is the winner. If the value of the goods is over £20, the person is disqualified. None of the contestants can see the prices on the articles.

1 Work out each contestant's total. Who won?

Tom Dickson
Iron, 2 packets pencils, kite, mirror and comb set

Marilyn Farrell
Hair curler, watch, ball, kite

Joan Andrews
Doll, jug, hair curler, screwdriver set

2 Find as many ways as you can of choosing goods to the value of exactly £20, as in the example.

Example	
1 packet pencils	£ 1
mirror, brush and comb	£ 5
iron	£ 8
doll	£ 6
Total	£20

148

Gift voucher

A big store advertised this special offer to attract customers.

Check these on your calculator:

| 2 | 0 | ÷ | 1 | 0 | = | ? |

| 4 | 2 | ÷ | 1 | 0 | = | ? |

| 9 | 8 | 5 | ÷ | 1 | 0 | = | ? |

FREE
GIFT VOUCHER
ONE POUND GIFT VOUCHER
FOR EVERY £10 YOU SPEND e.g.
SPEND £20 and receive a £2 GIFT VOUCHER
SPEND £42 and receive a £4 GIFT VOUCHER
SPEND £985 and receive a £98 GIFT VOUCHER

What value gift voucher would you receive for spending each of these amounts?

1 £50	**2** £80	**3** £90	**4** £100	**5** £150
6 £200	**7** £600	**8** £700	**9** £1000	**10** £1500
11 £750	**12** £680	**13** £890	**14** £410	**15** £990
16 £35	**17** £78	**18** £63	**19** £47	**20** £99
21 £12.99	**22** £45.50	**23** £23.99	**24** £119.95	**25** £239.50

What value gift voucher did each of these customers receive?

26 Mr Barnes, who bought an electric shaver.

£18.95

27 Mrs Campbell who bought a wall can opener and a kitchen tool set.

£6.99

£5.75

28 Derek Phillips, who bought this radio cassette player.

£49.99

29 Imram Jaffri, who bought this TV.

£199.99

30 Mr and Mrs Barton, who bought two garden chairs.

£19.99

SALE

31 Satinder Basra who bought a video camera.

£1049.00

32 Gloria Gordon, who bought these three pieces of jewellery.

33 John Appleby, who bought this suite.

Example

Mrs Donaldson received a voucher for £3. How much did she spend?

Answer: Mrs Donaldson spent between £30 and £39.99.

£3 *Gift Voucher*

---

34 Evan Jones received a voucher for £5. How much did he spend?

35 Brenda Hogg got a £7 gift voucher. How much did she spend?

36 Grace Millar got a £15 gift voucher. How much did she spend?

---

37 Mrs Suleman bought a patio set. She received a voucher for £10.
  (a) What did she spend?
  (b) How much of this was for chairs?
  (c) How many chairs were in the set?

SALE

SALE Chair £18.99

Table £29.99

Each of the customers below bought one or more of the items shown opposite. Work out what each customer bought.

38 Mrs Kaur – £2 voucher (One of her items was a toaster.)

39 Chu-Sin Wong – £8 voucher (She bought only 1 item.)

40 Mr Jenkins – £8 voucher (He bought 3 items and spent £1.60 less than Miss Wong.)

41 Mr Harris – £11 voucher (He bought 4 items.)

42 Mrs Tait – £3 voucher.

£14.99

£19.95

£9.99

£59.95

£16.45

£87.99

Buying money

Karen was going abroad on a school trip to France. Her dad gave her £50 spending money.

Karen found this table in the newspaper.

1 Write down the answers to Karen's 3 questions.

> What is French money called?

> How much would I get for £1?

> How much will I get for £50?

THE £

AUSTRIA	25.75 schillings	ITALY	2240 lire
BELGIUM	75.00 francs	MALTA	0.58 lira
CANADA	1.6625 dollars	NORWAY	10.81 kroner
DENMARK	13.39 kroner	PORTUGAL	200 escudos
EIRE	1.1950 punts	SPAIN	207 pesetas
FRANCE	11.30 francs	SWEDEN	10.67 krona
GERMANY	3.67 D-marks	SWITZERLAND	3.01 francs
GREECE	160 drachma	USA	1.27 dollars
HOLLAND	4.15 guilders	YUGOSLAVIA	268 dinars

2 While in Paris, Karen spent 126.90 fr on herself. She also bought 3 key rings, 4 model Eiffel Towers and 2 bottles of perfume as presents for her family.
How much did Karen spend altogether?
How much French money did Karen bring home?

3 Karen wanted to change her French francs back into pounds sterling.
How much did she get for them in British money? (Ask your teacher to help you.)
Remember:
  (i) Number of francs = number of £ × 11.30.
  (ii) Number of £ = number of francs ÷ 11.30.

4 (a) Change £70 into francs.
   (b) Change £125 into francs.
   (c) Change 500 francs into pounds.
   (d) Change 750 francs into pounds.

---

5 Dave changed £65 into francs. He spent 540 francs in Paris.
How much did he get for his francs when he got home?
(Go through the steps in questions 1 to 3 to solve this problem.)

6 (a) Change £5000 into American dollars.
   (b) Change £1500 into Spanish pesetas.
   (c) Change £750 into Greek drachmas.
   (d) Change £2000 into Italian lire.

---

7 (a) Change 3000 Deutschmarks into pounds sterling.
   (b) Change 25 000 Belgian francs into pounds sterling.
   (c) Change 850 Norwegian kroner into pounds sterling.
   (d) Change 50 000 Italian lire into pounds sterling.

8 Dave's dad changed £100 into dollars at this rate: | USA .... 1.27 dollars |

(a) How many dollars did he get?
Two weeks later he noticed this in the paper: | USA .... 1.23 dollars |

Dave's dad decided to change his dollars back into pounds.
(b) How much did he get?
(c) How much has he made?

Sports centre

Dave and Karen are both 16. Neither of them is a
member of the Sports Centre.
They go swimming together about 10 times a
year, and never swim for more than an hour.
Use the table opposite to answer these questions.

Bishopbridge Sports Centre		
Admission	Adults (over 18)	Juniors and OAPs
Members	Free	Free
Non-members	0.30	0.20
Swimming (per hour)	0.40	0.25
Membership (12 months)	Local residents	Others
Family	8.00	16.00
Adult (over 18)	3.50	7.00
Junior (under 18)	2.00	4.00
OAP	Free	Free

1 How much does Dave pay each time he goes
swimming?

2 How much does he pay per year?

3 How much does Karen pay each time she goes
swimming?

4 Karen is a local resident but Dave is not.
How much would each of them pay for 1 hour's
swimming if they became members of the
Sports Centre?

5 Both Dave and Karen are wondering how many
times in the year they would have to go swimming to
make it worth their while joining the Sports Centre.

(a) Copy this table and
fill it in.

Non-members

Number of visits	0	1	2	3		9	10
Dave and Karen each pay							

(b) Draw a graph to show how much Dave and Karen
each pay using your table.
Use $\frac{1}{2}$ cm squared paper for your graph.
Use the scales shown here.
Your graph should go up to 24 along the
bottom and £11 at the side.

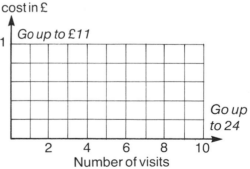

(c) Extend your graph to show the cost of up to 24 visits.

(d) Copy this table then fill it in.

Members

Number of visits	0	1	2	3		9	10
Dave pays							
Karen pays							

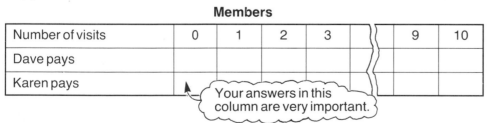

Your answers in this
column are very important.

(e) Draw two more lines on your graph to show
the information in your second table.
Extend the new lines to show the cost of up to
24 visits.
You can now help Dave and Karen. How many
times would they have to go swimming to make
joining the Sports Centre worthwhile?

Rent-a-room

The graph shows the rent and rates paid for a city-centre office.
Rent is paid to the owner of the building.
Rates are paid to the local council.

Example
An office has an area of 150 square feet.
How much was paid altogether in rent and rates in 1977?
How much of this was rent, and how much rates?

Total rent and rates $= 150 \times £4.32$
$\qquad\qquad\qquad\qquad = £648$

Rent $\quad = 150 \times £2.50$
$\qquad\quad = £375$

Rates $= £648 - £375$
$\qquad\quad = £273$

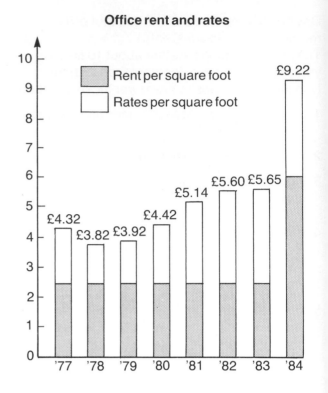

Office rent and rates

Rent per square foot
Rates per square foot

£9.22
£5.14
£5.60 £5.65
£4.32
£3.82 £3.92
£4.42

'77 '78 '79 '80 '81 '82 '83 '84

Calculate the total rent and rates paid for each office.

Also work out how much was paid in rent and how much in rates.

Set out your working as in the example.

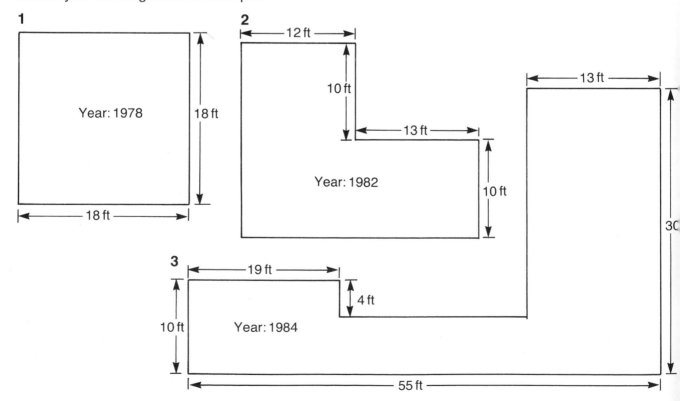

1

Year: 1978

18 ft

18 ft

2

12 ft

10 ft

13 ft

Year: 1982

10 ft

13 ft

30

3

19 ft

4 ft

10 ft

Year: 1984

55 ft

4 In 1982 a city office paid £2240
in rent and rates.
The office is square in shape.
  (a) What is the area of the office?
  (b) What is the length of the office?

5 Fillup Barlowe, private detective,
decided to open a small office in
the city at the beginning of 1984.

He had only rent to pay because
the local council was giving
new businesses one year free
of rates.

He worked out that he could afford
£720 rent for the year.

  (a) What is the area of the biggest
office he could afford?
  (b) Make scale drawings to show
two possible plans for his new
office. (Assume it was a simple
shape.)
Use the scale: 1 cm represents
1 foot.

6 (a) Work out the area of this office.
  (b) In which year did the rent plus
rates for this office go over
£6000?

Scale: 1 cm to 5 feet

7 Imagine your classroom is a
city office.
  (a) Find the area of the office.
  (b) Copy this table.
  (c) Fill in the table to show
what you would have
paid in each of the years.
  (d) Draw a graph like the one
on the previous page to
show all these figures.

	1977	1978	1979	1980	1981	1982	1983	1984
Rent								
Rates								
Total								

Number patterns

Study these 5 patterns carefully.

Answer questions 1, 2 and 3 *for each pattern*.

1 Draw Shape 1, Shape 2, Shape 3, Shape 4 and Shape 5.

2 Make a table like the one below. (This particular table is for Pattern 1. The table for Pattern 3 will read 'Number of spots' in the second row.)

Fill in your table up to shape 10. These clues will help you with the patterns.

Pattern 1 $3 \times 1, 3 \times 2, 3 \times 3 \ldots$
Pattern 2 $1 + 4, 2 + 4, 3 + 4 \ldots$
Pattern 3 $2 \times 1, 2 \times 2, 2 \times 3 \ldots$
Pattern 4 $2 \times 1 + 1,$
　　　　　$2 \times 2 + 1,$
　　　　　$2 \times 3 + 1 \ldots$
Pattern 5 $3 \times 1 - 2,$
　　　　　$3 \times 2 - 2,$
　　　　　$3 \times 3 - 2 \ldots$

	Shape 1	Shape 2	Shape 3	
Pattern 1				
Pattern 2				
Pattern 3				
Pattern 4				
Pattern 5				

Table for pattern 1

Shape	1	2	3	4		10
Number of matches	3	6	9			
Pattern	3×1	3×2	3×3			

3 How many matches or spots are there in the following
　(a) Shape 12
　(b) Shape 15
　(c) Shape 25
　(d) Shape 30
　(e) Shape 50
　(f) Shape 100?

4 How many matches or spots are there in Shape number *n*?
These clues will help you with this question. (The clues are not in the correct order.)

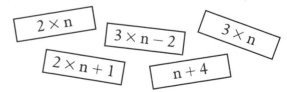

$2 \times n$　$3 \times n - 2$　$3 \times n$
$2 \times n + 1$　$n + 4$

5 Find a formula for Shape number *n* for each of these patterns.

Pattern 6

Pattern 7

Copy these sentences and fill in each blank with the correct number.

6 In Pattern 1 the biggest triangle you could make from 196 matches is Shape —. You would have — matches left over.

7 In Pattern 2 you would use 84 matches to make Shape —.

8 In Pattern 3 there are exactly 150 spots in Shape —.

9 In Pattern 5 you would use 58 matches to make Shape —.

10 In Pattern 6 there are 400 matches in Shape —.

11 In Pattern 7 there are 81 matches in Shape —.

12 Dave is helping arrange his brother's wedding reception at an hotel. Each ? stands for a different missing number. Work out all the missing numbers.

Here are some of our table plans, sir.

Table Plans

Table 1

Table 2

Table 3

If you have a Table 11 we could have everyone at the same table with every seat filled.

I'm sorry, sir, the biggest we have is a Table 8. But I could give you two Table ?s instead . . .

. . . or a Table 2 and a Table ?

Dave

If we had two Table ?s we could invite Uncle Bob and Aunt Madge with Bert and Penny. They might be upset if we missed them out.

Three Table ?s would do as well, or ? Table 2s.

Dave's brother

Football special

Study this advert carefully.

Now answer these questions:

1 Steve, Phil and Bob (all 15 years old) live in Stoke and support Manchester United. They see this advert and decide to go to the match with Steve's dad and his uncle.
   How much did the train fares cost for all of them?

2 Steve and his dad caught the early train. It left on time and arrived on time.
   How long did the journey take?

3 The other three fans caught the later train. It left on time but arrived 9 minutes late.
   How long did the journey take them?

4 Steve and his dad waited at the station in Manchester for the others to arrive.
   How long did they wait?

FOOTBALL SPECIAL

On Wednesday 12 December

European Cup winners Cup Man Utd. v. Juventus at Old Trafford

Stoke ....
Manchester Warwick Road .... dep. 1715 1750
                                          arr. 1804 1840
Returning from Warwick Road at 2200 or if extra time is played approximately 30 minutes later.

SECOND CLASS **£4** RETURN FARE

(Children 5 and under 16 years — £2.00)

BOOK EARLY – ACCOMMODATION LIMITED

In accordance with Bye-Law 3(a) of the British Railways Board's Bye-Laws, the Board give notice that the taking of intoxicating liquor onto all vehicles of the above trains is prohibited.
MAXIMUM PENALTY £200.

5 Copy what Steve said and fill in the two times.

We should be back in Stoke at about . . . . . or about . . . . . if there is extra time.

Timetable

Answer these questions using the timetable below.

1 At what time does the first train leave Southbay in the morning?
2 When does it arrive in Stanmore? How long does it take?
3 At how many stations does the 0730 stop between Southbay and Stanmore?
4 How far does the 0800 from Southbay go?
5 How long does the noon express take to travel to Stanmore? What other difference is there between this train and most of the others?

6 Mrs Jenkins lives in Southbay. She has a dental appointment in Stanmore at 2.00 pm. Which is the best train for her to catch?
7 Jim lives in Southbay and Carol lives in Westheath. They arrange to meet either on the train or in Stanmore at around 8 pm. Jim gets to Southbay station at 5.25 pm and takes the first train. Carol arrives at Westheath station at 6.35 pm. Explain what will happen.

Train departures from Southbay						
Southbay	**Balthorpe**	**Hamstone**	**Westheath**	**Goldacre**	**Bilton**	**Stanmore**
0730	0748	0803	0828	0906	0926	0954
0800	0818	0833				
0920	0938	0953	1018	1056	1116	1144
1110	1128	1143	1208	1246	1306	1334
1200	→	→	1240	→	→	1345
1235	→	→	1310			
1315	1333	1348	1413	1451	1511	1539
1400	1418	1433				
1520	→	→	1600			
			1608	→	→	1713
1715	→	→	→	→	→	1845
1730	1748	1803	1828	1906	1926	1954
1820	→	→	1900	→	→	2015
2015	2033	2048	2113	2151	2211	2239

Car ferry

Seaside Shipping

FERRY SERVICES

WESTPORT PIER — MILLHAVEN SLIP

LEAVING WESTPORT PIER
daily

0655$_{NS}$ 0715   0745
0815   0915   0945
and at 15 and 45 minutes
past each hour until
2015   2045$_F$
(except 1245 and 1645)

NS *Not Sundays*   F *Fridays only*

LEAVING MILLHAVEN SLIP
daily

0715$_{NS}$ 0730   0800
0830   0930
and on the hour and 30 minutes
past each hour until
2030   2100$_F$
(except 1300 and 1700)

Crossing takes only 10 minutes.

CHARGES FOR ACCOMPANIED CARS (excluding driver)

	Single	Day Return
Car (any length)	**£4.00**	**£6.10**
Driver/Passenger each	**£0.75**	—

Return fare is double the single fare

Millhaven Slip is situated 3 miles from Sandbay. There is a bus connection between the Ferry Slip and Sandbay – bus fare extra.

Bus leaves Sandbay 20 minutes before each sailing.

---

1 What is the normal interval between ferries from Westport?

2 When does the first ferry leave Westport (a) on Thursdays   (b) on Sundays?

3 When does the last ferry leave Millhaven (a) on Tuesdays   (b) on Fridays?

4 Mr and Mrs Fairlie arrive in Westport on a bus tour at 10.30 am one Tuesday morning. They decide to spend the day in Sandbay.

We leave Westport at 7 o'clock tonight.

(a) What is the earliest ferry they can take?
(b) How much will the ferry cost them? Give exact details.
(c) When is the latest they can leave Sandbay?
(d) Explain why the total spent on fares is more than your answer to (b).

5 Mr and Mrs Lennox and their two children arrive in Westport by car at 12.30 pm on a Bank Holiday Monday. They join the queue for the ferry.

This is the fourth ferry to sail since we arrived dad . . . and we are the last car to get on!

(a) At what time does their ferry sail?
(b) How much do their return ferry tickets cost them? Give exact details.
(c) They leave Millhaven by the second last ferry?
At what time does it sail?
(d) They queued for 45 minutes for the return ferry from Millhaven.
Estimate how long they spent in Sandbay.
(e) Counting their day as starting when they arrived in Westport and finishing when they left Millhaven, what fraction of their day was spent queuing?

Steam train

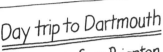

Day trip to Dartmouth

Steam train from Paignton to Kingswear, and ferry ride to olde-worlde Dartmouth for tea.

(Check time of last train back from Kingswear, and leave Dartmouth by ferry at least half an hour before.)

Answer these questions using the timetable opposite.

1 Easter Monday was 8th April in 1985. On which other Monday during April and May 1985 was the train running?

2 What does the sign ≋ beside Paignton mean?

3 At what time does the 10.40 train from Paignton arrive in Kingswear?

4 If you did not get off the 10.40 train but went straight back to Paignton, when would you arrive back at Paignton?

TORBAY & DARTMOUTH LINE 1985
5 to 11, 14, 17, 21, 24, 28 April; 1, 5, 6, 8, 12, 15, 19, 22 May; then daily
26 May to 22 September; 25, 29 September, 2, 6, 9, 13, 16, 20, 23, 27 October.

STANDARD SERVICE
(including Saturdays and Sundays during Peak Season)

B.R. Arrivals	10.28	11.53	14.12	16.20
	P	P	P	P
Paignton ≋	10.40	12.05	14.30	16.30
Goodrington	10.45	12.10	14.35	16.35
Churston	10.55	12.20	14.45	16.45
Kingswear (for Dartmouth)	11.10	12.35	15.00	17.00
Kingswear	11.20	12.45	15.15	17.15
Churston	11.35	13.00	15.30	17.30
Goodrington	11.45	13.10	15.40	17.40
Paignton ≋	11.50	13.15	15.45	17.45

THE NATION'S HOLIDAY ROUTE

P = Pullman First Class Observation Car runs in this train (Supplementary fare.)

GO ANY TRAIN–RETURN ANY TRAIN

PEAK SERVICE: 22 July to 30 August—Mondays to Fridays

Paignton ≋	—	10.20	11.05	11.50	12.35	14.05	14.50	15.35	16.20	17.00
Goodrington	—	10.25	11.10	11.55	12.40	14.10	14.55	15.40	16.25	—
Churston	09.50	10.35	11.20	12.05	12.50	14.20	15.05	15.50	16.35	17.15
Kingswear (for Dartmouth)	10.05	10.50	11.35	12.20	13.05	14.35	15.20	16.05	16.50	—
Kingswear	10.15	11.00	11.45	12.30	14.00	14.45	15.30	16.15	17.15	—
Churston	10.35	11.20	12.05	12.50	14.20	15.00	15.50	16.35	17.30	—
Goodrington	10.45	11.30	12.15	13.00	14.30	15.15	16.00	16.45	17.40	—
Paignton ≋	10.50	11.35	12.20	13.05	14.35	15.20	16.05	16.50	17.45	—

5 You arrive in Paignton at 12.15 on 12th May. How long do you have to wait for the train to Kingswear?

6 You arrive in Paignton at 14.30 on 25th April. How long do you have to wait for the train?

7 The Alston family arrive in Paignton at 10.45 on 30th July 1985.
 (a) What is the first train they can take to Kingswear?
 (b) When will they arrive in Kingswear?
 (c) They take the ferry to Dartmouth and plan to return by the last train. When should they leave Dartmouth (at the latest)?

 (d) For how long altogether were they on the steam train that day?

8 Can the standard service be run using only one locomotive? Explain your answer.

9 (a) The peak service needs two locomotives running at the same time. Explain how you can tell this from the timetable.
 (b) The railway is a single track. There must be a place with a double track where they can pass each other. Where is this place? Explain from the timetable.
 (c) Where do you think the two engines are kept during the night in summer?

The rate for the job

Clive is a grade 3 mechanic.

Mechanic Grade 3	Basic rate: £4 per hour
	Overtime: time and a half
	Saturday work: double time

Example

Basic rate is £4 per hour. What are time-and-a-half and double time?

Time and a half is £4 per hour $+ \frac{1}{2}$ of £4 per hour
= £4 per hour + £2 per hour
= £6 per hour

Double time is $2 \times$ £4 per hour
= £8 per hour.

Work out how much Clive earned during each of the following weeks.

1 Week beginning 2 March
35 hours basic
4 hours overtime
3 hours on Saturday

2 Week beginning 9 March
35 hours basic
6 hours overtime
$1\frac{1}{2}$ hours on Saturday

3 Week beginning 16 March
35 hours basic
$3\frac{1}{2}$ hours overtime
4 hours on Saturday

4 Week beginning 23 March
35 hours basic
$6\frac{1}{2}$ hours overtime

5 Week beginning 13 April
35 hours basic
No overtime

6 Week beginning 20 April
35 hours basic
$5\frac{1}{2}$ hours overtime

7 The maximum overtime allowed is 6 hours on week nights and 4 hours on Saturday. What is Clive's maximum wage?

8 What is the maximum wage of a grade 1 mechanic, whose basic rate is £5 per hour?

9 For the week beginning 30 March Clive earned £170.
He did not work on Saturday.
How many hours overtime did he work that week?

10 For the week beginning 6 April Clive earned £170. He worked twice as many hours of weekday overtime as he did on Saturday. How many hours weekday overtime did he work and how many hours on Saturday?

11 Copy this table. Fill in each wage rate.

	Basic rate	**Time and a half**	**Double time**
Storeman Grade 2	£2 per hour		
Storeman Grade 1	£3 per hour		
Mechanic Grade 3	£4 per hour		
Mechanic Grade 2	£4.50 per hour		
Mechanic Grade 1	£5 per hour		
Supervisor	£5.40 per hour		

```
        Weekday rates

0800-1600   basic rate
(one-hour lunch break)
1600-1630   tea-break
1630 onwards   time and
a half
        Saturday rate

All Saturday work
double time
Starting time 0900
```

Copy each employee's time sheet.
Work out their wages for the week.

> TIMEKEEPING
>
> Employees are allowed a total of 15 minutes lateness per week.
> If that total is exceeded, ALL lateness is penalised each day in units of 15 mins, e.g. 3 mins late = 15 mins pay docked at basic rate.

12

EMPLOYEE Colin Green	GRADE Mechanic Grade 2	
Week beginning 12/5/86		
	Start	Finish	
MON	0805	1600	7
TUES	0800	1930	7 + 3
WED	0800	1600	7
THU	0745	1830	7 + 2
FRI	0815	1600	7
SAT	—	—	0
Basic 35 hours at £ 4.50		£
Time & ½ 5 hours at £ 6.75		£
Double 0 hours at £ 9.00		£
TOTAL		£
Deduction for latecoming		£
PAY		£

13

EMPLOYEE Angela Carr	GRADE Mechanic Grade 1	
Week beginning 12/5/86		
	Start	Finish	
MON	0755	1600	
TUES	0803	1830	
WED	0758	1600	
THU	0804	2000	
FRI	0759	1600	
SAT	0900	1200	
Basic hours at £		£
Time & ½ hours at £		£
Double hours at £		£
TOTAL		£
Deduction for latecoming		£
PAY		£

14

EMPLOYEE Tom Daly	GRADE Storeman Grade 1	
Week beginning 19/5/86		
	Start	Finish	
MON	0806	1600	
TUES	0809	1700	
WED	0810	1600	
THU	0805	1730	
FRI	0817	1600	
SAT	—	—	
Basic hours at £		£
Time & ½ hours at £		£
Double hours at £		£
TOTAL		£
Deduction for latecoming		£
PAY		£

---

15 In the week beginning 16 June 1986, the firm had a special contract to finish. A team of employees had to work this overtime:

Tuesday 17 June '86: 1630 – 1930
Thursday 19 June '86: 1630 – 2100
Saturday 21 June '86: 0900 – 1230

Here is the team who worked overtime on each occasion.

Work out the total amount the firm paid in overtime that week to finish the contract.

16 The firm also pays a completion bonus of £25 to each worker if the contract is finished on time. This contract *was* completed. How much did each member of the team earn that week?

Clive Jones
Mechanic Grade 3

Angela Carr
Mechanic Grade 1

Colin Green
Mechanic Grade 2

Fred Spanner
Supervisor

Happy days

1 Teachers are paid on the last Thursday of each month.
Miss Primrose is paid £8500 per annum.
(a) Make a list of the dates on which she was paid for the year shown in the calendar.
(b) What amount was on each pay cheque (to the nearest penny)?

2 The authorities were thinking of changing the teachers' pay-day to the Thursday before the last Friday of each month. What difference would this have made to Miss Primrose in the year shown in the calendar?

3 Some teachers wanted to be paid every fourth Thursday, starting on 27th January.
(a) Make a list of pay-days under this new system.
(b) What amount would have been on each of Miss Primrose's pay cheques?
(c) Why do you think some teachers wanted this system rather than either of the other two?

1983

JANUARY					
M	3	10	17	24	31
T	4	11	18	25	
W	5	12	19	26	
T	6	13	20	27	
F	7	14	21	28	
S	1	8	15	22	29
S	2	9	16	23	30

FEBRUARY				
M	7	14	21	28
T	1	8	15	22
W	2	9	16	23
T	3	10	17	24
F	4	11	18	25
S	5	12	19	26
S	6	13	20	27

MARCH					
M	7	14	21	28	
T	1	8	15	22	29
W	2	9	16	23	30
T	3	10	17	24	31
F	4	11	18	25	
S	5	12	19	26	
S	6	13	20	27	

APRIL				
M	4	11	18	25
T	5	12	19	26
W	6	13	20	27
T	7	14	21	28
F	1	8	15	22	29
S	2	9	16	23	30
S	3	10	17	24

MAY					
M	2	9	16	23	30
T	3	10	17	24	31
W	4	11	18	25	
T	5	12	19	26	
F	6	13	20	27	
S	7	14	21	28	
S	1	8	15	22	29

JUNE					
M	6	13	20	27	
T	7	14	21	28	
W	1	8	15	22	29
T	2	9	16	23	30
F	3	10	17	24	
S	4	11	18	25	
S	5	12	19	26	

JULY				
M	4	11	18	25
T	5	12	19	26
W	6	13	20	27
T	7	14	21	28
F	1	8	15	22	29
S	2	9	16	23	30
S	3	10	17	24	31

AUGUST				
M	1	8	15	22	29
T	2	9	16	23	30
W	3	10	17	24	31
T	4	11	18	25
F	5	12	19	26
S	6	13	20	27
S	7	14	21	28

SEPTEMBER				
M	5	12	19	26
T	6	13	20	27
W	7	14	21	28
T	1	8	15	22	29
F	2	9	16	23	30
S	3	10	17	24
S	4	11	18	25

OCTOBER					
M	3	10	17	24	31
T	4	11	18	25	
W	5	12	19	26	
T	6	13	20	27	
F	7	14	21	28	
S	1	8	15	22	29
S	2	9	16	23	30

NOVEMBER				
M	7	14	21	28
T	1	8	15	22	29
W	2	9	16	23	30
T	3	10	17	24
F	4	11	18	25
S	5	12	19	26
S	6	13	20	27

DECEMBER				
M	5	12	19	26
T	6	13	20	27
W	7	14	21	28
T	1	8	15	22	29
F	2	9	16	23	30
S	3	10	17	24	31
S	4	11	18	25

Think about it

1 The Youth Club meets every Saturday night. The activities are as shown on the right: Make up an events diary for the Youth Club for January, February and March 1988. The first Saturday in 1988 is 2 January.

Disco	first and third Saturdays of the month
Games	second Saturday of the month
Swimming	last Saturday of the month

2 Choose an activity for the blank week. When does a month have 5 Saturdays? Give as much detail as you can.

3 Why are Christmas day and New Year's day always on the same day of the week? (i.e. if Christmas day is Thursday, New Year's day is Thursday.)

4 Study these clues carefully.
(a) What are the two months?
(b) Explain the information in clue 2 and clue 3.
(c) In which years does month A have an extra day?

1 Month B follows immediately after month A.

2 Month B usually starts on the same day of the week as month A.

3 Every so often month B begins one day later in the week than month A.

Christmas day

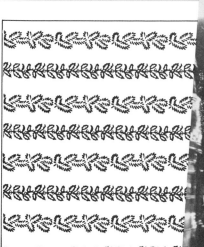

This table shows on which day of the week Christmas day fell between 1975 and 1985.

Christmas day	
1975	Thursday
1976	Saturday
1977	Sunday
1978	Monday
1979	Tuesday
1980	Thursday
1981	Friday
1982	Saturday
1983	Sunday
1984	Tuesday
1985	Wednesday

1 Copy the table.

2 Use the pattern in the table to continue it up to 1995.

3 Explain the pattern in the table.
Here are some hints to help you:

Hints
How many days are there in a year (usually)?
How many days are there in a leap year?
How many days are there in a week?
Why would Christmas fall on the same day of the week every year if there were 364 days in a year?

4 On which days of the week will your birthday fall for the next 10 years?

Delivery dates

This instruction is from a mail-order firm.
Use the calendar on the opposite page to answer these questions.

Allow 28 days for delivery

1 When should you expect delivery of goods ordered on the following dates:
(a) 4 May   (b) 3 February   (c) 25 March   (d) 13 June
(e) 8 July   (f) 11 September   (g) 17 November   (h) 26 October?

2 Do you see a quick way of doing these without counting on 28 days?
Use the quick way to find delivery dates for the following order dates:
(a) 8 February   (b) 10 March   (c) 21 April   (d) 23 April
(e) 19 August   (f) 23 August   (g) 1 October   (h) 14 December?

Speed limits

Example
Jean's mum walked 1 mile in 10 minutes.
What was her speed in miles per hour? (mph)

Draw two number lines and extend them to one hour (60 minutes).

Jean's mum walked 6 miles in 60 minutes. She walked at 6 mph.

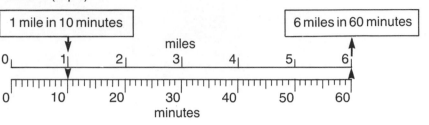

---

1 Mr Smith drove 3 miles along a main road in a built-up area. The journey took him 6 minutes.
Did he break the speed limit or not?
Copy these two number lines.

Continue the pattern to find Mr Smith's speed.

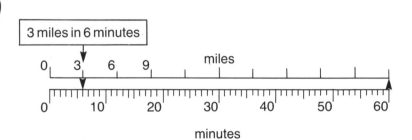

Work out the speed of each of the following journeys.
The speeds will be in either miles per hour (mph) or kilometres per hour (kmph).

2 Joe walked 3 miles in 20 minutes.

3 Karen's dad drove 6 miles in 10 minutes. Did he break the speed limit?

4 A bus driver drove 5 miles in 15 minutes.

5 Mr Haynes drove 25 km in France in 30 minutes.
Did he break the speed limit?

6 Pierre drove 13 km in 12 minutes.

7 An athlete ran 1 mile in 4 minutes.

8 A sprinter ran 100 metres in 10 seconds.

9 A train covered a 1 mile stretch of track in 1 minute.

---

10 This diagram shows Moira's route home from work yesterday.
Work out her speed for each section of the journey and say whether she kept to the speed limits or not.

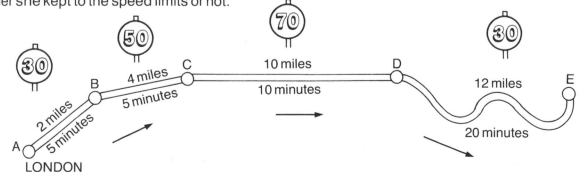

Example

Mrs Bond drives a distance of 5 km in 8 minutes. What was her speed?

Mrs Bond's speed was between 35 kmph and 40 kmph. (It looks like 37.5 kmph from the number line.)

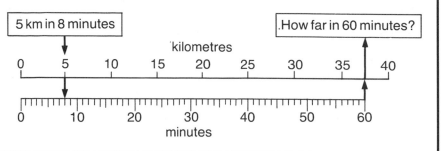

Work out approximate speeds for each of these journeys.

11 A journey of 12 km in 14 minutes.

12 A journey of 8 miles in 9 minutes.

13 A journey of 10 km in 7 minutes.

14 A journey of 12 km in 8 minutes.

15 This diagram shows M. Lebon's route home from work each evening. His time for each section yesterday is marked on the diagram. Work out his speed for each section as accurately as you can and say whether he kept within the speed limits or not.

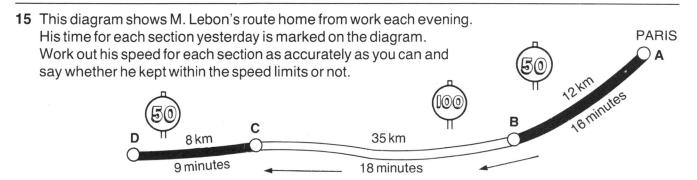

A speed formula

This formula tells you how to work out an average speed.

$$S = D \div T$$

S stands for the average speed on a journey

D stands for the distance travelled

T stands for the time taken

Example

A car travelled 45 miles in 1 hour 12 minutes. Work out the average speed of the car in mph. (mph means 'miles per hour'.)

D = 45 miles
T = 1 hr 12 mins
  = 1.2 hrs
S = D ÷ T
  = 45 ÷ 1.2
  = 37.5

Average speed was 37.5 mph.

minutes	hours	minutes	hours
6	0.1	1	0.02
12	**0.2**	2	0.03
18	0.3	3	0.05
24	0.4	4	0.07
30	0.5	5	0.08
36	0.6		
42	0.7		
48	0.8		
54	0.9		

1 Davinder travelled 5 miles in 20 minutes on his bicycle. Work out the average speed for the journey (in mph).

Hint From the table,
20 mins = 18 mins + 2 mins
        = 0.3 hrs + 0.03 hrs
        = 0.33 hrs

2 Use the formula to work out accurate answers for questions 11 to 15 above.

Travelling salesman

Ahmed is a travelling salesman.
On Monday he leaves home at 10 am, makes calls in two other towns,
Bridgeford and Porchester, then travels home.
This travel graph gives details of his journeys.

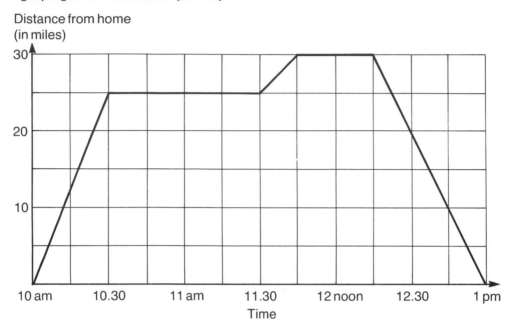

1 At what time did Ahmed make his first call?

2 How far is Bridgeford from Ahmed's home?

3 What was his average speed for this part of
his journey?

4 How long did he stay in Bridgeford?

5 At what time did he leave Bridgeford?

6 At what time did he arrive in Porchester?

7 How far is Porchester from Bridgeford?

8 What was his average speed for the journey
from Bridgeford to Porchester?

9 How long did he stay in Porchester?

10 How long did his journey home take?

11 What was his average speed for the journey
home?

12 Draw a travel graph like the one above to show
Ahmed's journeys on Tuesday morning.
Leave home 10.15 am, arrive in Bridgeford
10.45 am, leave Bridgeford 11 am, arrive in
Porchester 11.15 am, leave Porchester
12 noon, arrive home 12.45 pm.

13 Draw a graph for Ahmed's journeys on
Wednesday morning. Leave home 10 am,
held up in traffic 15 miles from home from
10.15 am to 10.30 am, drive straight to
Porchester arriving at 11 am, leave
Porchester 11.30 am, arrive Bridgeford
11.45 am, leave Bridgeford 12.15 pm,
arrive home 1 pm.

14 Draw a graph for Ahmed's journeys on
Thursday morning. Leave home 10 am,
travel to Porchester at average speed of
60 mph, stay in Porchester $1\frac{1}{4}$ hours, travel
back to Bridgeford at average speed of
20 mph, stay in Bridgeford $\frac{3}{4}$ hour, travel
home at average speed of 50 mph.

167

Two trains

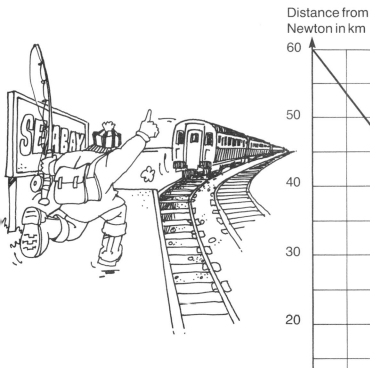

Distance from Newton in km (vertical axis, labelled 0, 10, 20, 30, 40, 50, 60)

Graph labels: (a), (b), (c), (d), (e), (f), (g), (h), (i)

Time (horizontal axis: 1400, 1410, 1420, 1430, 1440, 1450)

The travel graph opposite shows the journeys of two trains.
Train 1 is the 1400 from Seabay to Newton.
Train 2 is the 1405 from Newton to Seabay.

Answer these questions from the graph.

1 Which line tells you about which train?

2 Make out a table like this for each train. ▶

3 At what time did the trains pass each other?

4 How far were they from Newton when they passed?

5 How far apart are Newton and Seabay?

Section of journey	(a)	(b)	
Times	1405 to 1425 20 mins	1425 to 1440 15 mins	
Distance travelled	20 km	30 km	
Average speed	60 km/hr	120 km/hr	

6 The two tables below give details of two other trains that afternoon.

Train 3 departs Newton 1500 and arrives Seabay 1545.

Times	1500 to 1520	1520 to 1530	1530 to 1545
Average speed	90 km/hr	stopped	120 km/hr

Train 4 departs Seabay 1510 and arrives Newton 1555.

Times	1510 to 1540	1540 to 1545	1545 to 1555
Average speed	80 km/hr	stopped	120 km/hr

Draw a graph like the one above to show these two journeys.
Draw both lines on the same grid.

Would you credit that?

Look carefully at this advert. ▶

1 Copy and complete this calculation to check if the advert is true.

Cash price of cleaner	£85.99
Deposit	£ 9.49
9 payments of £8.50	
(9 × £8.50)	£
Total paid	£

(a) Is this total the same as the cash price?

(b) How much money do you actually hand over on the day you buy the cleaner?

(c) How long will it be before your debt is cleared?

2 Check each of the prices in this table as you did in question 1. ▶
(At least one of them has a mistake which you have to correct.)

3 You can buy a 3 piece suite for a cash price of £399 or put down a deposit and make 9 monthly payments of £39.50.
How much of a deposit do you need?

4 You can buy a carpet for £273.45 cash or put down a deposit of £25.95 then make 9 monthly payments.
How much will each monthly payment be?

Usually when you pay on credit you end up paying a bit more than the cash price.

5 Copy and complete this calculation to check the advert.

Cash price of TV	£299.99
Deposit	£ 29.95
24 payments of £14.06	£
Total paid	£
Cash price	£299.99
Credit charge	£

Another term for paying on credit is hire purchase (HP).

6 Work out the total HP price of each of these. ▶
Say how much extra you pay in each case.

9 months to pay at cash price

Example:
Vacuum cleaner
Deposit £9.49
9 monthly payments of £8.50

9 months to pay at cash price

Item	Cash price	Deposit	Monthly payment
(a) TV	£245.99	£25.49	£24.50
(b) Golf clubs	£75	£12	£7
(c) China	£119.95	£16.45	£11.50
(d) Dishwasher	£239.95	£38.45	£23.50

SONY 2060 COLOUR TV

2 way speaker system, teak finish. SPECIAL PURCHASE

£299.99

OR DEPOSIT £29.95, 24 MONTHLY PAYMENTS OF £14.06 TOTAL EXTENDED CREDIT PRICE £367.39   A.P.R. 24.8%

Item	Cash price	Deposit	Monthly payment
(a) Bicycle	£119.99	£12	12 @ £10.08
(b) Stereo	£249.99	£24.99	24 @ £11.50
(c) Stereo	£249.99	£24.99	36 @ £ 8.50
(d) Microwave oven	£350	£35	12 @ £29.40

Monthly payments

Linda saw this advert for a car in the newspaper.

How is the 'weekly equivalent' calculated?

Hints

How much do you pay in one year at £90 per month?

How much per week is this?

Why is this not the same answer as Linda got on her calculator?

Explain what you have discovered to your teacher.

| Monthly payments | £90 |
| Weekly equivalent | £20.77 |

There is something here I don't understand!

$9\ 0 \div 4 = 22.5$

Calculate the weekly equivalent for each of these.
Give answers to the nearest penny.

1 £110 per month
2 £45 per month
3 £96 per month

4 £200 per month
5 £126 per month
6 £68 per month

7 £74 per month
8 £86 per month
9 £184 per month

Work out the maximum monthly payment each man can afford to make.
Give answers to the nearest pound.

10 Mr Dale estimates that he can afford £35 out of his weekly wage to buy a car.

11 Mr Thomson can manage only £20 per week to buy a car.

This is part of a real advert for a car:

Check all the figures in the advert.

Explain to your teacher where these figures come from.
(a) £4006
(b) £3205
(c) £3589.56
(d) £99.71
(e) £23.01
(f) £4390.56

MICRA 1.1	
Cash price	£3846
On the road charge	£ 160	← This is the same for all cars
Total price	£4006	
Deposit	£ 801	← This is $\frac{1}{5}$ of the total price (ignore pence)
Balance to pay	£3205	
Finance charges	£ 384.56	← You are told this each time
Total to pay	£3589.56	
Monthly payment 36 × £	99.71	
WEEKLY EQUIVALENT £	23.01	
Total credit price	£4390.56	

Write down the complete list of figures for each of these cars.

12 CHERRY 1.3	Cash price	£4996.00
	Finance charges	£ 494.83
13 SUNNY 1.3	Cash price	£5066.00
	Finance charges	£ 501.55
14 BLUEBIRD 1.8	Cash price	£6189.00
	Finance charges	£ 609.42

More bills

Look carefully at this gas bill:
The formula in the box below shows how the bill was worked out.
Go through the formula, starting with the first two numbers in the bill.
Look carefully to see where the other numbers in the formula come from.

DATE OF READING	METER READING			GAS SUPPLIED		VAT %	CHARGES £
	PRESENT	*	PREVIOUS *	Cubic Feet (Hundreds)	THERMS		
12 SEP	3920		3845	75	78.000	0	27.46
STANDING CHARGE						0	9.90
				35.2 pence per therm			
				TOTAL AMOUNT DUE ▶ £			37.36

Total bill = £ (present reading − previous reading) × 1.04 × 35.2 ÷ 100 + £9.90

1 Calculate the gas bill due on each meter.

	Meter A	Meter B	Meter C	Meter D	Meter E
Present reading	1570	3472	8659	5754	6105
Previous reading	1320	3405	8463	5059	4975

2 Make out a complete bill for meter E.
Use the same headings as in the bill above.

3 Here are the readings on Mr Robb's gas meter.
The meter is read by the Gas Board every three months.

Meter reading

Mar '85	6372
Jun '85	6704
Sep '85	6787
Dec '85	7043
Mar '86	7546

(a) Work out Mr Robb's bills for June '85, September '85, December '85 and March '86.

(b) What was his total bill for the year?

(c) Mr Robb is told that he can pay for his gas in equal monthly payments
(i.e. 12 payments per year) in future.
The Gas Board usually makes these payments a whole number of pounds or a sum ending in 50p
(e.g. £12 or £12.50).
How much per month should Mr Robb pay for the next year?

(d) Mr Robb did make 12 payments over the next year.
Each payment was the amount you worked out in part (c).
How much did he pay that year?

(e) Here are the actual meter readings for the year that Mr Robb paid in monthly instalments.
Work out the total bill for gas used between March '86 and March '87.

Meter reading		Cost per therm	Standing charge
Jun '86	7842	35.2 p	£ 9.90
Sep '86	7966	35.2 p	£ 9.90
Dec '86	8140	37.5 p	£10.20
Mar '87	8596	37.5 p	£10.20

Notice the price increases for the third and fourth bills. These will change the formula.

(f) Mr Robb has already paid for his gas up to March '87. (See your answer to part (d)).
Does he owe the gas board money, or should he get a rebate?
If so, how much?

4 Mrs Arthur is an old age pensioner.
To save money, she hardly ever uses her
gas fire.
Her gas bill in September 1986 was £19.78.
Her meter reading in June 1986 was:

What was her meter reading in September
1986?

5 How many meter units give a gas bill of just
under the £100 mark when the standing
charge is £10.20?

Example
Here is how to check your electricity bill.

Present reading   41865
Previous reading   40828
Units used   1037

First 34 units @ 14.99p each   = £ 5.09
Remainder 1003 @ 4.73p each   = £47.44
Total amount   = £52.53

The Electricity Board ignores fractions of a penny
in the last column.

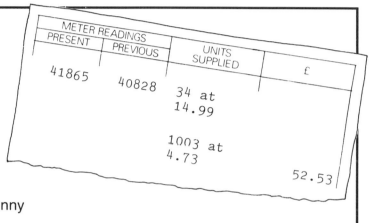

METER READINGS		UNITS SUPPLIED	£
PRESENT	PREVIOUS		
41865	40828	34 at 14.99	
		1003 at 4.73	
			52.53

---

6 Work out the electricity bill due on each meter.

	Meter A	Meter B	Meter C	Meter D
Present reading	32170	17516	21715	48529
Previous reading	32130	17451	20854	47529

7 In December 1986 Mrs Kidd's meter read: **4 4 5 3 8**

The reading in October 1986 was: **4 2 9 5 7**

She was out when the meter reader called.
Her bill in December looked like this:
(a) By how many units have they overestimated or
    underestimated?
(b) How much of a difference does this make to
    her bill?

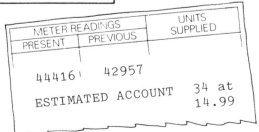

METER READINGS		UNITS SUPPLIED
PRESENT	PREVIOUS	
44416	42957	
ESTIMATED ACCOUNT		34 at 14.99

Taxi!

This table shows how taxi fares are worked out:

In the daytime:
as soon as you get into the taxi the meter reads   £00.80

as soon as you have gone $\frac{3}{5}$ of a mile the meter reads   £00.90

	Daytime	Between midnight and 6 am	Booked by phone (daytime)
First $\frac{3}{5}$ mile	80p	120p	100p
Every $\frac{1}{5}$ mile thereafter	10p	15p	12p
Waiting time	10p per minute		

This chart shows how much you are charged for the first 2 miles. Copy the chart and extend it to show fares up to 5 miles.

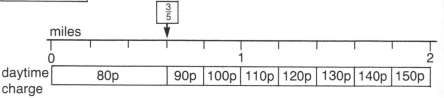

1 Work out the daytime fare for a journey of
   (a) 1 mile   (d) 5 miles   (g) 20 miles
   (b) 2 miles   (e) 10 miles   (h) 23 miles
   (c) 3 miles   (f) 15 miles   (i) 25 miles
   Explain why the fare for 10 miles is not double the fare for 5 miles.

2 Draw a chart like the one above to show taxi fares between midnight and 6 am. Make your chart go up to 6 miles.

3 Work out the fare for each of these journeys.
   (a) 1 mile at 3 am   (b) 3 miles at 0500 hours
   (c) 5 miles at   (d) 10 miles at one o'clock
       5.30 am   in the morning.

4 Draw a chart to show taxi fares when you book by phone.
   Show distances up to 5 miles as before.

5 Work out the cost of each of these daytime phone bookings:
   (a) 1 mile   (b) 3 miles   (c) 5 miles
   (d) 14 miles   (e) 20 miles   (f) 25 miles

The chart at the bottom of the page will help you with questions 6 and 7.

6 Mrs Gomez hailed a taxi in the street at 3.15 pm.
   It took her to the shops, waited for 17 minutes, then took her back home.
   The distance travelled was 2.8 miles.
   How much was the taxi fare?

7 Steve stopped a taxi outside the disco at 1 am. When he got home (4.1 miles away) he discovered he had spent all his money and forgotten his key.
   It took him 13 minutes to get his brother to come to the door and let him in so he could find money to pay the taxi driver.
   How much did he have to pay?

Circle city

This is a plan of Circle City.
Each road is named after the
direction it faces.

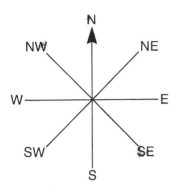

If you stand at the centre of the
city and look along North Road,
this is the view you see.

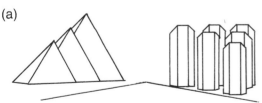

1 Which roads do you look along to see each of the views below?

(a)

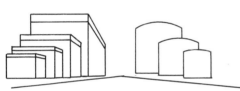

(d)

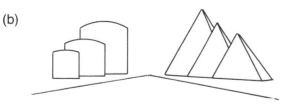

(b)

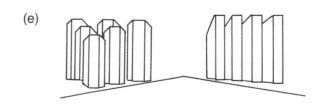

(e)

(c)

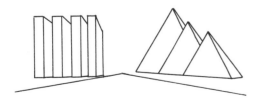

(f)

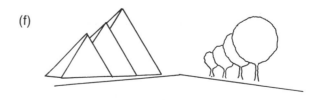

2 Sketch the view along the North-West Road.

Finding your bearings

This compass shows many directions.
Each direction is measured in degrees.

Another name for a direction is a **bearing**.

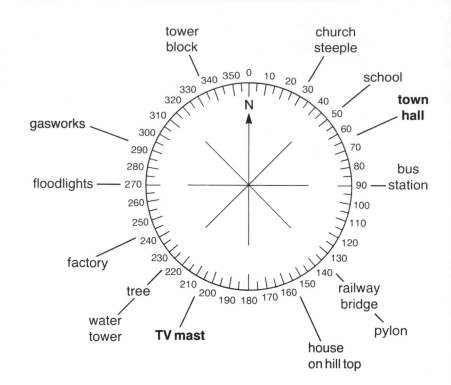

> ### Example 1
> In which direction would you look to see the town hall?
>
> In direction 065°.
>
> ### Example 2
> What is the bearing of the TV mast?
>
> In direction 205°.

Answer these questions:

1 In which direction would you look to see the school?

2 What is the bearing of the gasworks?

3 What is the bearing of the bus station?

4 What do you see if you look in direction 030°?

5 Which two objects are on a bearing of 225°?

6 What is on a bearing of 090°?

7 What is the bearing of the railway bridge?

8 What other object is on the same bearing as the railway bridge?

9 What is the bearing of the floodlights?

10 Copy this table and fill it in.

Direction	N	S	E	W	NE	SW	NW	SE
Bearing		180°						

11 What lies in the opposite direction to the house on the hill top?
Copy this table and fill in the blanks.

		Bearing
Object in opposite direction →	House	
	....	
	Difference in bearings	

12 Make up tables like the one in question 11 for the bus station and the town hall.

13 Use a real compass to find the bearings of landmarks you can see from the classroom window.

14 Take your compass to an open area and find a landmark for each of the 8 main points of the compass.

Orienteering

This is the plan of an orienteering course.

Competitors run round the course carrying a compass.

Scale:
1 cm represents 100 m

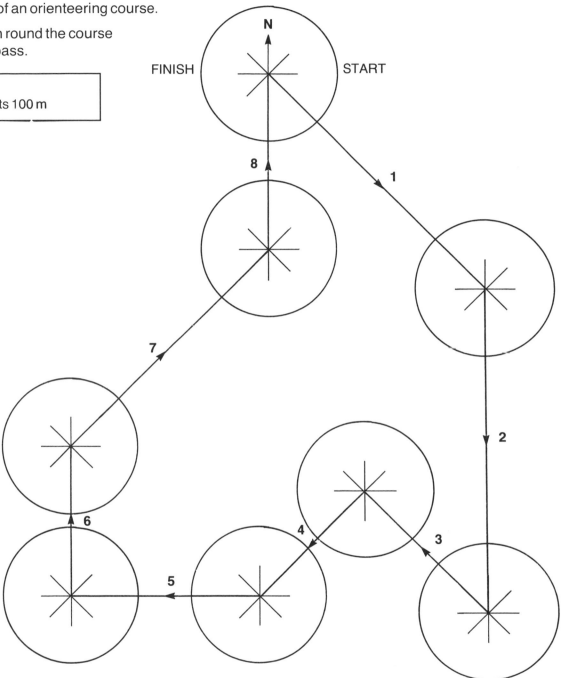

FINISH   START

1 Copy this table and complete it for each section of the course. You will need a protractor to measure each angle.

Section	1	2
Bearing	SE (135°)	S (180°)
Distance	800 m	850 m

2 Calculate the total distance covered by a runner who completes the course.

Speedboat

1 Copy the navigator's table below.
Fill it in.

Section	Red buoy to green buoy	Green buoy to yellow buoy	Yellow buoy to red buoy
Course	°	°	°
Distance	km	km	km

Speedboat Time Trials
Course 1

Scale:
1 cm represents 1 km

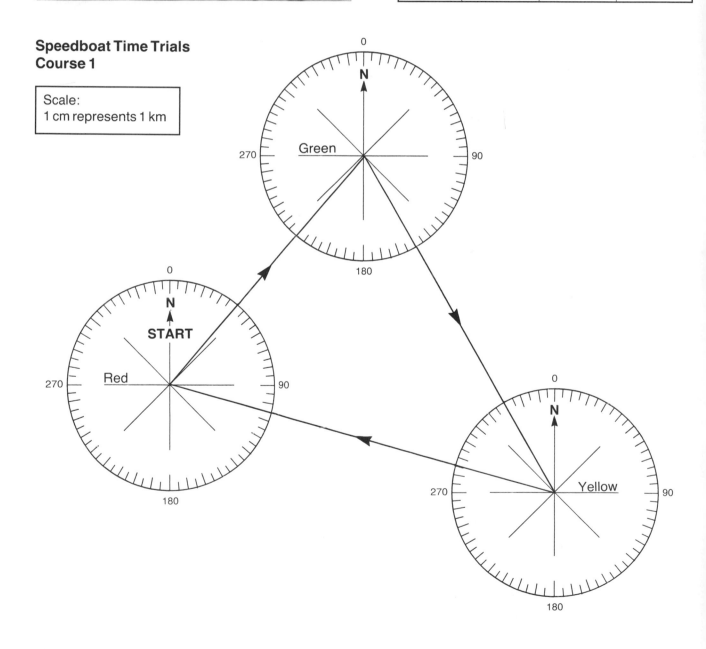

2 Make out the navigator's table for Course 2.

Speedboat Time Trials
Course 2

Scale:
1 cm represents 1 km

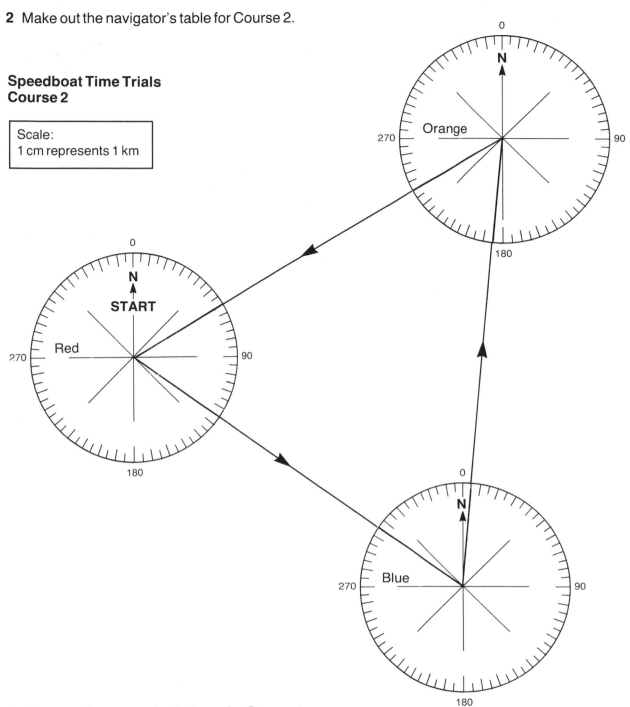

3 Here are the competitor's times for Course 1:

Dupont	Kellermann	Macdonald	Briggs
17 mins 08 secs	17 mins 12 secs	16 mins 59 secs	17 mins 00 secs

Here are the times for Course 2:

Dupont	Kellermann	Macdonald	Briggs
20 mins 12 secs	20 mins 09 secs	20 mins 10 secs	20 mins 13 secs

Add each competitor's two times together.
Who came 1st, 2nd, 3rd and 4th?

Equilateral triangle

1 (a) Trace this triangle into your notebook.
Remember to trace the shaded part and the spot.
(b) Trace the triangle again on to tracing paper.
(c) Fit the tracing paper over the triangle in your notebook.
(d) Stick a pin through both spots.
(e) Turn the triangle to look like these two diagrams.

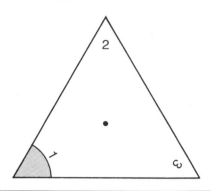

2 What did you discover about the size of the 3 angles of the triangle?

3 What is the total of the 3 angles of the triangle in degrees?

4 What size is each angle of this triangle?
A triangle like this is called an **equilateral triangle**.

In each of these diagrams, the shapes are either squares or equilateral triangles.
Write down the size of each shaded angle.
Do not measure the angles.

5
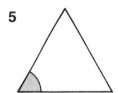

6
Middle point

7
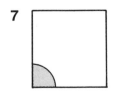

8

Check your answers to questions 5-8 with your teacher before you go on.

Write down the size of each shaded angle.
Do not measure the angles.

9 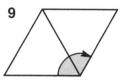   **10**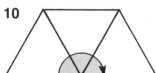

11   **12**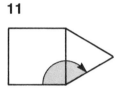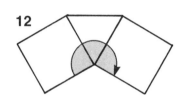

Here is how to draw an equilateral triangle using only a ruler and compasses.

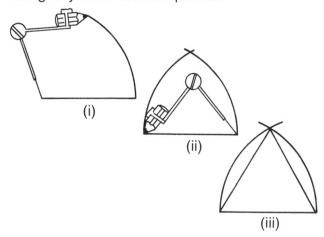

(i)

(ii)

(iii)

13 Using only a ruler and compasses, draw angles of these sizes:

(a) 60° (b) 45° (c) 120°

(d) 150° (e) 135° (f) 75°

Make the drawings on squared paper.
(Some of the diagrams for questions 4-13 will help you with these problems.)

One hundred per cent

> **Example**
> People often talk about 100 per cent. The headline opposite quotes a football manager before an important game. Remember that 100% means the whole lot. What do you think the headline means?
>
> ## Robson demands 100 per cent
> England manager Bobby Robson said yesterday that he expected '100% effort' from each player in his world cup squad. Although quietly confident, he clearly feels England will not succeed without the total commitment

Look at the following examples.
Write down what you think each one means.

1

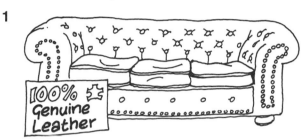

5

You won't play on Saturday unless you are 100% fit!

2

I got 100% in the maths exam.

6 Carl Lewis won 4 gold medals at the 1984 Olympic Games.
Explain in your own words what he said about running in the Olympics.

> Lewis had said that sprinting in the Olympics was 1% mental and 99% physical beforehand. But on the day: "It's 100% mental."

3

I am not quite 100% certain that I know who committed this crime.

7 The Olympic gold medal is actually silver coated with 6.5 grams of 24-carat gold.
It is worth about £90.
The silver and bronze medals are exactly what you would expect – 100% pure.

4 Explain exactly what this special offer is.

100% OFF THIS HEADBOARD when you buy one of our beds

Sale

1 Who do you think is right?   **2** Work out what each person should pay.

All dressed up

Mr and Mrs Spruce are all dressed up for their wedding.
The labels below are from their clothes.

Mr Spruce

Tie

| 100% POLYESTER |

Shirt

| 65% POLYESTER |
| 35% COTTON |

Underwear

| 100% COTTON |

Suit

| 55% POLYESTER |
| ? % WOOL |

Socks

| 51% LAMBSWOOL |
| ? % NYLON |

Mrs Spruce

Blouse

| 60% COTTON |
| ? % POLYESTER |

Jacket

| 45% WOOL |
| ? % POLYESTER |
| 20% ACRYLIC |
| 15% MIXED FIBRES |

Underwear

| 65% POLYESTER |
| ? % NYLON |
| 5% ELASTANE |

Skirt

| 100% COTTON |

Tights

| 100% NYLON |

This diagram shows what
Mr Spruce's shirt is made of.

```
0  10 20 30 40 50 60 70 80 90 100%
|-----------------|--------|
|    Polyester    | Cotton |  Shirt
```

Draw a diagram like this for
each of the items of clothing
mentioned above.

Percentages and fractions

Example

50% means 50 out of every 100.

25% means 25 out of every 100.

20% means 20 out of every 100.

10% means 10 out of every 100.

1 (a) What percentage of this shape is shaded?

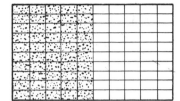

(b) What fraction of this shape is shaded?

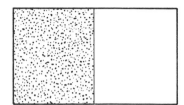

Do the same for the next 3 questions. Write down the percentage of shape (a) that is shaded, and the fraction of shape (b) that is shaded.

2 (a)

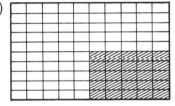

(b)

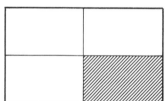

3 (a)

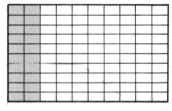

(b)

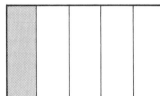

4 (a)

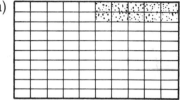

(b)

5 Use your answers to questions 1 to 4 to fill in this table.

Memorise this table.

Percentage of shape		Fraction of shape
50%	is the same as	
25%	is the same as	
20%	is the same as	
10%	is the same as	

6 (a) 68 pupils sat the maths exam. The pass rate was 50%.
How many passed the exam?

(b) 8000 fans attended a concert. 25% of them paid at the door. The rest had tickets.
How many paid at the door?
How many had tickets?

7 How many extra ml of washing up liquid are there in the larger bottle?

8 Copy this shape.

Shade 50% of the shape red,
25% of the shape blue,
10% of the shape yellow.
What percentage of the shape is unshaded?

182

Choose the answer

The answer to each of these questions is either 50%, 25%, 20% or
10% (or very close to one of these).
Choose the correct answer for each one.

1 What percentage of this shape is shaded?

3 What percentage of this
plank is being cut off?

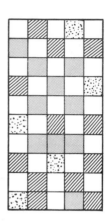

2 The price of this car is to increase by £470 next
week.
What percentage increase is this?

4 What percentage of
these squares is
shaded like this ☐ ?
What percentage of
these squares is
shaded like this ▦ ?
What percentage of
these squares is
shaded like this ▨ ?

5 Here are the results of a survey of the people
who live in a street.
(a) How many people live in this street?
(b) What percentage of the people are between
20 and 49 years old?
(c) What percentage are in their twenties?
(d) What percentage are in their fifties?
(e) What percentage are under 10 years of age?
(f) Here is part of the survey in more detail.
What percentage of the people living in the
street are teenagers?

Age (years)	Tally
0 – 9	IIII IIII IIII III
10 – 19	IIII IIII IIII IIII IIII IIII I
20 – 29	IIII IIII IIII IIII IIII IIII IIII IIII
30 – 39	IIII IIII IIII IIII IIII I
40 – 49	IIII IIII IIII IIII IIII IIII IIII
50 – 59	IIII IIII IIII IIII IIII IIII IIII IIII II
60 and over	IIII IIII

Age (years)	Tally
10	III
11	IIII
12	IIII

6 (a) What percentage of this rectangle
is shaded ☐ ?
(b) What % is shaded ▨ ?
(c) What % is shaded ▦ ?
(d) What % is shaded ☐ ?

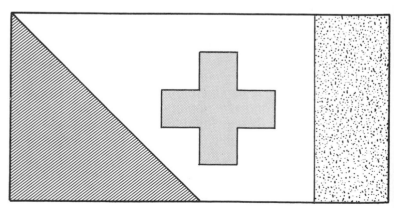

Granted

The Council have agreed to give grants for house repairs.

1 (a) What percentage of the repair bill will the Council pay?
  (b) What percentage will each householder have to pay?

2 Work out how much each householder in this table will pay.

3 What total amount will the Council pay out on these 6 homes?

4 The Council is forced to make cuts before these repairs are agreed.
  (a) What percentage will each householder now pay?
  (b) Work out the amount each householder will pay at the new rate.
  (c) What total amount will the Council now pay out on the 6 homes?
  (d) How much has the Council saved by cutting the grant from 90% to 75%?

Householder	Repairs	Cost
Mr Jackson	Roof retiled	£2000
Mrs Simpson	Roof retiled, new windows	£3500
Mr Jenkins	New gutters	£ 950
Mr Ali	Roof retiled, new gutters	£3100
Mary Grant	Windows replaced	£2875
Jim Groves	New roof, gutters	£3259

Getting a loan

Example

Sharon and Richard want to buy a house costing £24 000.
They ask a Building Society for a loan.
The Building Society offers them a 90% loan.
How much do they need for a deposit?

Loan is 90% of £24 000
Deposit is 10% of £24 000
  $= \frac{1}{10}$ of £24 000
  $= £2400$
Sharon and Richard need £2400 for a deposit.

1 Work out the deposit needed for each of these 90% loans.
  (a) £28 000 (b) £35 000 (c) £36 500 (d) £40 000
  (e) £45 000 (f) £47 750 (g) £52 900 (h) £154 000

2 Work out the deposit needed for loans of 80% on each price.
  (a) £25 000 (b) £30 000 (c) £28 000 (d) £26 500
  (e) £32 000 (f) £45 500 (g) £58 000 (h) £37 550

Surveys

Example

A survey team asked 600 shoppers this question: *'Which brand of margarine do you buy?'* The chart on the right shows the results of the survey.

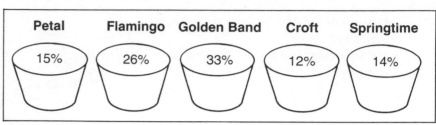

Petal	Flamingo	Golden Band	Croft	Springtime
15%	26%	33%	12%	14%

How many shoppers bought Petal margarine?
15% means 15 out of every 100.
15% of 600 means:

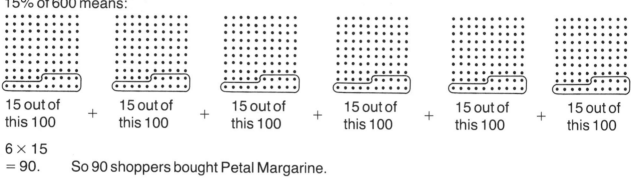

15 out of this 100 + 15 out of this 100 + 15 out of this 100 + 15 out of this 100 + 15 out of this 100 + 15 out of this 100

6×15
$= 90.$ So 90 shoppers bought Petal Margarine.

1 Work out the number of shoppers who bought each of the other brands.

T 2 Draw a piechart to show all this information. (Your teacher will give you a percentage piechart.)

3 A survey team asked 500 dog owners this question: *'Which brand of dog food does your dog prefer?'* The chart on the right shows the results of the ▶ survey.

(a) How many dogs preferred each type of dog food?

(b) Do you think this advert for Chunky is telling the truth or not? ▶

(c) Draw a piechart to show the results of the survey.

4 Here are the results of a survey of washing up liquids:

(a) How many housewives prefer each brand of washing up liquid? (1500 housewives were interviewed in the survey.)

(b) Make up an advert for Palmsoft, something like the advert for Chunky.

(c) Draw a piechart to show the results of the survey.

VAT

Example
Check this bill using your calculator.
15% means 15p in every £
15% of £30 is 30 × 15p
= 450p
= £4.50

Building Centre	
40 slabs @ 75p each	£30
VAT at 15%	£ 4.50
TOTAL	£34.50

Make out bills for each of these customers.
Remember to add VAT to each bill.

1

5 tins of vinyl matt emulsion, please.

Building Centre	
5 tins paint @ £5.40	
VAT at 15%	
TOTAL	

3

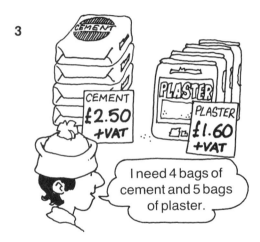

I need 4 bags of cement and 5 bags of plaster.

2

I want 9 sections of fencing and 10 posts.

FENCING
Posts – £6.50
Panels – £12.00

Building Centre	
9 fence panels @ £12 each	£
10 posts @ £6.50 each	£
Subtotal	£
VAT at 15%	£
TOTAL	£

4

I want 10 sheets of hardboard.

Mr Andrews does not have a van, so he will have to have the hardboard delivered.
Which store would give him the better price?
(Show all your working.)

Fit the shapes

1 Copy each of these shapes
   on to cardboard as follows:
   (i) Trace the shape.
   (ii) Transfer to cardboard –
        use a pin to make holes
        at the corners, then join
        up the pin holes.
   (iii) Cut out the shape.
   (iv) Name each of the
        shapes as precisely as
        you can.

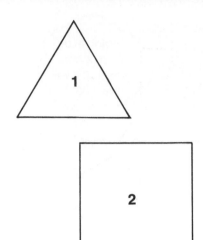
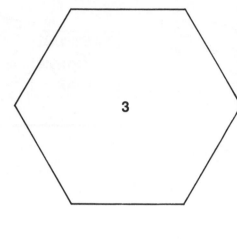

2 Copy each of these tilings using your cardboard shapes as
   templates.
   (a) Each tile in this tiling is
       Shape 1.
   (b) Each tile in this tiling is
       Shape 2.
   (c) Each tile in this tiling is
       Shape 3.

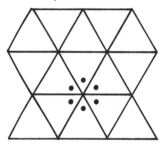
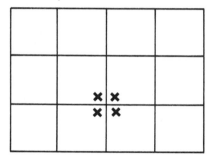
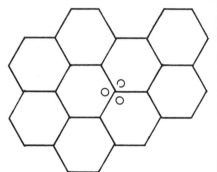

3 For each tiling in question 2,
   find the total of the marked
   angles.
   What is special about this
   answer?

4 What is special about the
   answer to this sum?

   $90° + 120° + 90° + 60°$

   What is the connection
   between this drawing and
   the sum?
   Discuss this with your
   teacher.

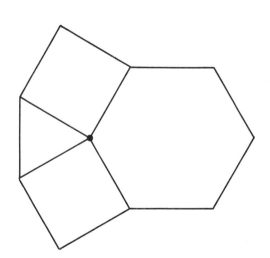

5 Copy the pattern below on to a large sheet of paper using your
cardboard shapes as templates.
(Your shapes will be larger than those drawn here.)

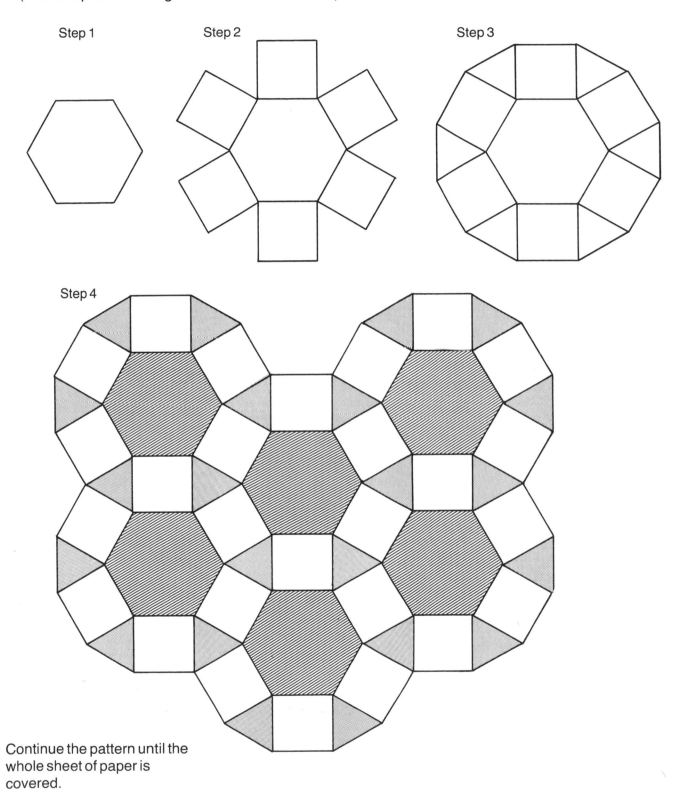

Step 1   Step 2   Step 3

Step 4

Continue the pattern until the
whole sheet of paper is
covered.

6 Choose any point where 4 lines meet in this pattern.
  (a) Write down the sizes of the 4 angles at this point.
  (b) What is the total of the 4 angles?

Fit the cooker

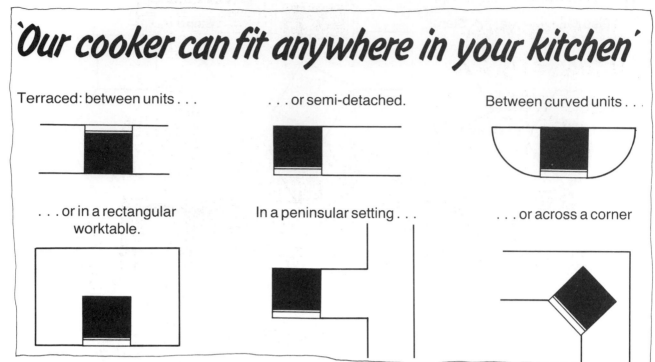

'Our cooker can fit anywhere in your kitchen'

Terraced: between units . . .

. . . or semi-detached.

Between curved units . . .

. . . or in a rectangular worktable.

In a peninsular setting . . .

. . . or across a corner

The top of the cooker is 600 mm square.
Make accurate scale drawings to show how the cooker fits into the 6 positions shown in the diagrams.
Use the scale 1:20. (This means 1 mm represents 20 mm.)

A tiling problem

1 Copy each of these shapes on to card.

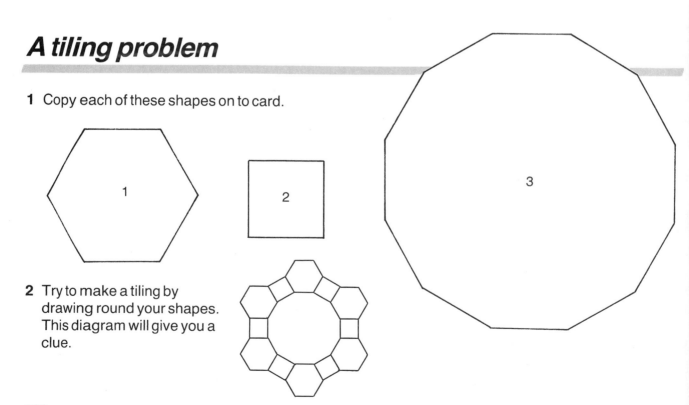

1

2

3

2 Try to make a tiling by drawing round your shapes. This diagram will give you a clue.

Rising prices

Here is how Tony worked out the new prices.

Earphones   5% means 5p in every £
            So 5% of £8 is 8 × 5p
                        = 40p
            New price   = £8 + 40p
                        = £8.40

1 Work out the new price of each item in the picture.

2 Work out the new price of these other items in the shop.

Item	Present price
Computer	£399
Hi-fi	£750
TV	£299
Tape cleaner	£9

Computer
5% means 5p in every £
So 5% of £399 is 399 × 5p
            = ... p
            = £. . . . .
New price   = £399 + £. . . . .
            = £. . . . . .

3 The prices of some items in the shop have to increase by 8%.
Work out the new price of each of these.

Item	Present price
Video	£450
Vacuum cleaner	£175
Shaver	£25
Washing machine	£165

Video
8% means 8p in every £
So 8% of £450 is 450 × 8p
            = ... p
            = £. . . . .
New price   = £450 + £. . . . .
            = £. . . . .

Changing shapes

1 (a) Draw a rectangle with an area 35% more than this one.

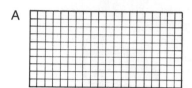

A

Area of rectangle A = 20 × 10 = 200 squares. 35% means 35 in every 100.
35% of 200 squares = 2 × 35 squares
$\qquad$ = 70 squares
New area is:
200 squares + 70 squares = 270 squares
This rectangle has an area 35% more than rectangle A:

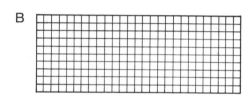

B

(b) Dave drew this rectangle as his answer to question 1 (a).

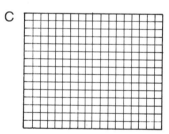

C

Should he be marked right or not?

Draw your answers to parts
(c) to (f) on 2 mm squared paper.

(c) Draw a rectangle with an area 60% more than rectangle A.
(d) Draw a rectangle with an area 75% more than rectangle A.
(e) Draw a rectangle with an area 20% less than rectangle A.
(f) Draw a square with an area 50% less than rectangle A.

2 (a) Draw a rectangle with an area 15% more than this square.

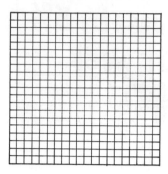

(b) Draw a rectangle with an area 40% more than this square.
(c) Draw a rectangle with an area 65% more than this square.
(d) Draw a rectangle with an area 30% less than this square.
(e) Draw a square with an area 19% less than this square.

3 Make 3 copies of this shape on 5 mm squared paper.

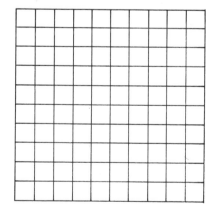

(a) Show how to reduce the area of the shape by 30% by making one straight cut.
(b) Show how to reduce the area of the shape by 49% by making two straight cuts.
(c) Show how to reduce the area of the shape by 18% by making one straight cut.

A little extra

1 (a) How many cubes make up this cuboid?

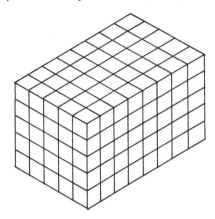

(b) How many cubes would there be in a cuboid which was 30% bigger than the one in question 1 (a)?

2 Which of these cuboids could show the answer to question 1(b)?

A

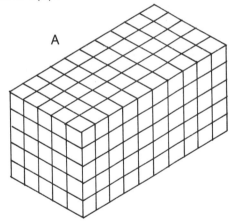

B

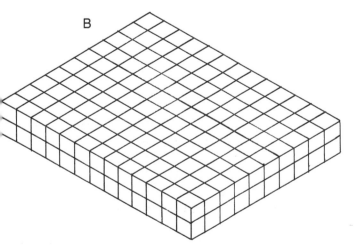

3 This tin is full of cocoa.
(a) How many cm³ of cocoa does the tin hold?

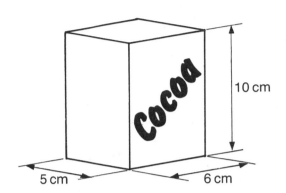

10 cm

5 cm   6 cm

(b) How many cm³ of cocoa will this new larger tin hold.

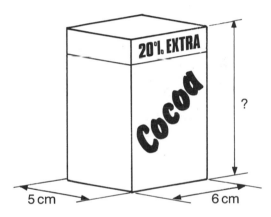

20°/₀ EXTRA

?

5 cm   6 cm

(c) Design a tin to hold the increased amount and mark in the dimensions.
(d) Draw a half-size net of the new tin.

4 A chocolate bar measures 10 cm × 5 cm × 2 cm.
(a) How many cm³ of chocolate is this?
(b) The makers want to reduce the amount of chocolate by 10% but keep the price the same.
How many cm³ will there be in the new bar?
(c) What size should they make the new bar?

5 A box of talcum powder measures 20 cm × 8 cm × 5 cm.
(a) How many cm³ is this?
(b) The makers want to design a presentation box in the shape of a cube which contains 25% more than the other box.
What size should the cube be?

Wage negotiations

Round 1

We want a 15% wage rise.

You will have to work 15% extra hours and increase production by 15%

Not likely! We already work 40 hours a week and produce 2000 cars per week . . . and you pay us only £120 a week!

Round 2

We want to install new machinery in the factory . . . and cut the work force by 12%.

If you agree to this we will give you an 8% wage rise. You will increase production by 10% . . . and we will *reduce* your hours by 5%.

1 Copy this table, then fill it in. You must do percentage calculations to find the answers for the second and third rows.
You will have to think very carefully how to work out the answers in the 'Hours per week' column.

	Wages per week (£)	Production per week (cars)	Hours per week	Workforce
Before negotiations				150
Round 1 proposals				
Round 2 proposals				

2 Would you accept the round 2 proposal?

Hints

15% means 15 out of 100

15% means . . . . . out of 10

15% means . . . . . out of 40

What is 15% of 40 hours?

5% means 5 out of 100

5% means . . . . . out of 10

5% means . . . . . out of 40

What is 5% of 40 hours?

3 Every grade of worker in the factory got a rise of 8%.
Here are the wages before the rise:

Copy this table and fill in all the blanks.

	Labourer	Semi-skilled	Skilled	Foreman	Manager
Wage (per week)					
Rise					
New wage					
Difference between wage and manager's wage — Before					—
Difference between wage and manager's wage — After					—

Suppose this had been the result of the wage negotiations:

4 Copy the table above again and fill in all the blanks for a £15 increase.

5 Who would be better off and who would be worse off with the £15 compared with the 8% increase?

Reminder
Count 1 year as 52 weeks.

194

Smashing reductions

Window prices smashed!

1 You measure and fit — SAVE 50%

2 We measure, you fit — SAVE 40%

3 We measure and fit — SAVE 25%

Example

Mr Barr needed a window listed at £130.
The firm measured the window but he fitted it himself.
How much did he pay?

List price	= £130
Saving	= 40% (which is 40p in every £)
Saving on £1	= 40p
Saving on £130	= 130 × 40p
	= 5200p
	= £52
Price paid	= £130 − £52
	= £78

---

1 Copy this table, then fill it in.
Do your working as in the example above.

Customer measures	Firm measures	Customer fits	Firm fits	% saving	List price	Saving	Price paid
	✓	✓		40%	£130	£52	£78
	✓		✓		£ 80		
✓		✓			£150		
	✓		✓		£ 78		
	✓	✓			£184		
	✓		✓		£200		
✓		✓			£172		

2 For each of these windows, who measured and who fitted?

(a) Mr Frame

I got this for £43.50

LIST PRICE £87

(b) Mrs W. Payne

I only paid £180 for this window!

LIST PRICE £240

Very interesting

Karen got £50 for her 16th birthday.
She decided to put it in the bank.

Here is the first page of her new bank book: ▶

High Street Bank Interest rate 8% p.a.

Date	Code	Deposit	Payment	Balance
22/6/86	CASH	50.00		50.00

She got another £50 for her 17th birthday.
She went back to the bank.
Here is how the computer marked up her bank
book: ▶

Date	Code	Deposit	Payment	Balance
22/6/86	CASH	50.00		50.00
	INT	4.00		54.00
22/6/87	CASH	50.00		104.00

Where did that extra £4 come from?

The £4 is your interest for the year.

p.a. is short for 'per annum' and means each year.

8% p.a. means 8p in every £
8% of £50 is 50 × 8p
= 400p
= £4

1 Work out how Karen's book will look after she
puts in her 18th birthday present of £50.
Remember that the interest will be 8% of £104.

2 Suppose Karen had put all her money into the
Commercial Bank where the interest rate is
12% p.a.
Show how her book would have looked after
her 18th birthday.

3

3 Cleaners WIN £60 000

Mrs Donnelly

I will put my share in the bank and get interest at 9%.

Mrs Woods

I will buy this car and buy a flat with the remainder.

Mrs Woods sold both her car and her flat one
year later.
The price she got for the car was 30% less
than she paid for it. Her flat was worth 15% more.

I will put £12 000 in the bank at 9% per annum. And I would like to buy these two paintings.

Mrs Bowie

Mrs Bowie's paintings both increased in value
over the next year. The dearer one increased in
value by 6% and the cheaper one by 12%. She
sold them both.

How much was each of the three ladies' shares
worth after one year?
Whose share do you think had been put to best
use?

Bathroom tiles

Example

Look at this design for a bathroom tile.

Give it a quarter turn clockwise and it looks like this:

Give the first tile a half turn and it looks like this:

Give the first tile a three-quarters turn and it looks like this:

These are the only four ways you can stick this tile on the wall.

1 If you put them all on the same way you get this pattern:

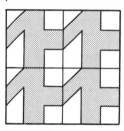

Copy the pattern.

2 If you put them on using the four different ways you get this pattern:

Copy the pattern.

3 If you put them on using only half turns you get this pattern:

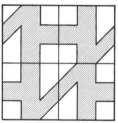

Copy the pattern.

Do you see why the pattern on each tile *must* join up?

4 (a) Draw the four different positions for this tile.

(b) Cover a large area with tiles all facing the same way.

(c) Cover a large area with tiles facing all four ways.

5 Repeat question 4 for these patterns:

(a)

(b)

(c)

(d)

(e)

(f)

6 Design 2 new tiles like the previous ones.
Draw this shape on $\frac{1}{2}$ cm squared paper as a guide. Your design must go through all the dots.

7 This pattern is made from nine identical tiles all facing the same way.
(The dividing lines between tiles are not shown.)
(a) Draw one of the tiles.
(b) Draw the pattern formed when the tiles face four ways.

8 This pattern is made from 16 identical tiles facing four ways.
(The dividing lines between tiles are not shown.)
Draw the four different positions of the tile.

9 Each of these patterns is made from 16 identical tiles using half turns.
(a) Draw one of each tile.

(b) Draw the patterns formed when the tiles face 4 ways.

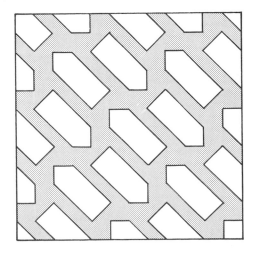

Flexitables

Flexitables are intended for use at conferences and business meetings.
Here are the three shapes of table:

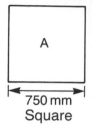

750 mm
Square

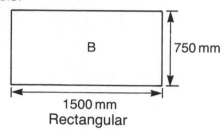

750 mm

1500 mm
Rectangular

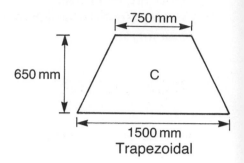

750 mm
650 mm

1500 mm
Trapezoidal

1 (a) Make a scale drawing of each shape on cardboard.
 Use the scale 1 : 20. (This means that 1 mm on your drawing represents 20 mm on the real table.)
 (b) Cut out each shape and show the shapes to your teacher.

---

2 Here is one way of making a solid conference table to seat 8.

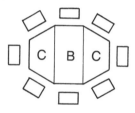

 (a) Draw this table using your cardboard shapes.
 (b) How would you make a solid conference table to seat 6, using only 2 tables?

3 (a) How would you make this solid conference table? (Use only type C tables.)

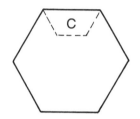

 How many does it seat?
 (b) If the table could be hollow (have a space in the middle), how many type C tables would you need?

4 (a) How would you make this solid table? Draw a diagram.

 (b) It looks as if it might seat 9. Explain why this might not be very comfortable.
 (c) How would you make a solid triangular table which would seat 15 comfortably?

5

Would you make out a table plan for 18, please, Miss Jones. It's for next month's conference.

	Price list
Table A	£ 89 each
Table B	£142 each
Table C	£142 each
Chairs	£ 48 each

Miss Jones came up with 2 ideas for the table.
She put in a report to her boss with drawings and prices.
Do the same yourself. Here are some hints: ▶

Plan 1
Solid
One type only needed

Plan 2
Hollow
All 3 types needed

6 Measure the angles of a type C table.
Write the size of each angle on your diagram
for question 4.
Write down anything you notice.

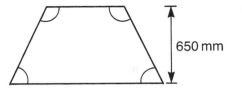

650 mm

7 Why is the width of a type C table only 650 mm?

Triangle puzzle

(i)   Copy each of these triangles on to cardboard.
(ii) Cut out each one.
(iii) Write the size of each angle on the card.

1 One of the triangles is a
right-angled triangle.
Which one?

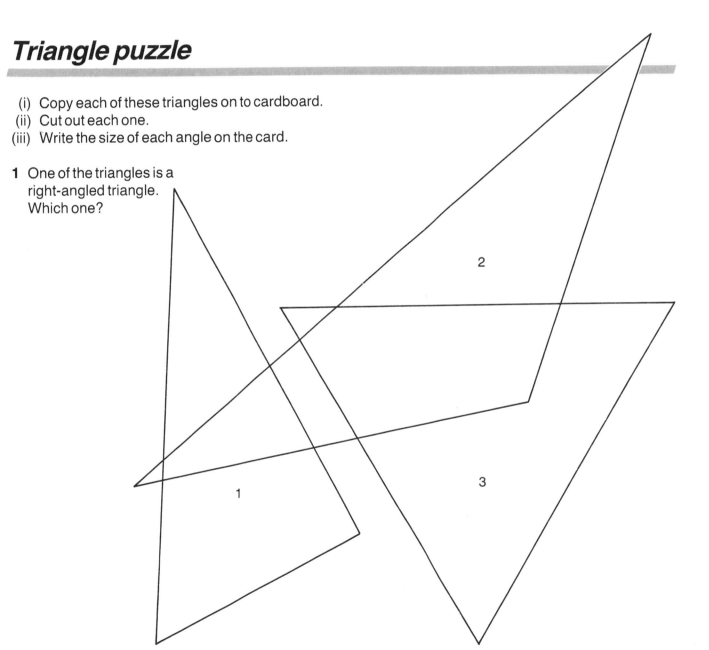

2 Fit two of the triangles together to make one right-angled triangle.
   (a) Draw round your shapes to show how you solved the problem.
   (b) Write the size of each angle on the new triangle.

3 Fit together all three of your triangles to make another right-angled
   triangle.
   (a) Show how you solved the problem as in question 2.
   (b) Write the size of each angle on the new triangle.

Suite dreams

Lisa and Paul are getting married. Lisa's mum is giving them £1000 as a wedding present.
They saw this advert in the evening paper and decided to spend their £1000 at this store. They made a list of the furniture they would like.

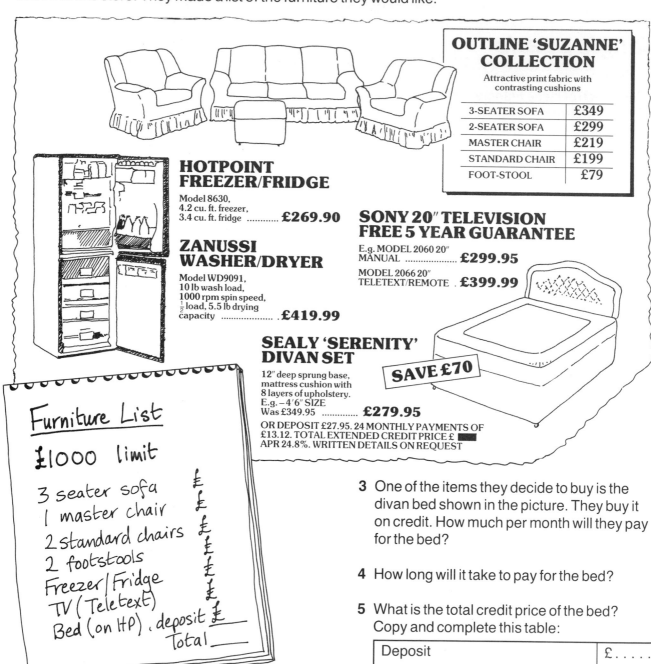

OUTLINE 'SUZANNE' COLLECTION

Attractive print fabric with contrasting cushions

3-SEATER SOFA	£349
2-SEATER SOFA	£299
MASTER CHAIR	£219
STANDARD CHAIR	£199
FOOT-STOOL	£79

HOTPOINT FREEZER/FRIDGE

Model 8630,
4.2 cu. ft. freezer,
3.4 cu. ft. fridge **£269.90**

ZANUSSI WASHER/DRYER

Model WD9091,
10 lb wash load,
1000 rpm spin speed,
½ load, 5.5 lb drying
capacity **£419.99**

SONY 20″ TELEVISION FREE 5 YEAR GUARANTEE

E.g. MODEL 2060 20″
MANUAL **£299.95**

MODEL 2066 20″
TELETEXT/REMOTE . **£399.99**

SEALY 'SERENITY' DIVAN SET

12″ deep sprung base,
mattress cushion with
8 layers of upholstery.
E.g. – 4'6″ SIZE
Was £349.95 .... **£279.95**

SAVE £70

OR DEPOSIT £27.95. 24 MONTHLY PAYMENTS OF £13.12. TOTAL EXTENDED CREDIT PRICE £ ■■■ APR 24.8%. WRITTEN DETAILS ON REQUEST

Furniture List

£1000 limit

3 seater sofa   £
1 master chair   £
2 standard chairs   £
2 footstools   £
Freezer/Fridge   £
TV (Teletext)   £
Bed (on HP) . deposit £ ____
Total ____

1 Copy Lisa and Paul's list and work out the total.

2 If they do not have enough money for all that is on the list, what should they leave out? Make out a new list which they could afford.

3 One of the items they decide to buy is the divan bed shown in the picture. They buy it on credit. How much per month will they pay for the bed?

4 How long will it take to pay for the bed?

5 What is the total credit price of the bed? Copy and complete this table:

Deposit	£ . . . . .
24 payments of £	£ . . . . .
Total credit price	£ . . . . .

6 Explain the label 'Save £70' on the bed.

7 How much will Lisa and Paul save on the original price of the bed if they buy on credit?

Leisure centre

This advert is for a supermarket's range of garden furniture. ▶

Tony is opening an outdoor leisure centre.
He ordered this furniture from the supermarket:
10   patio tables
10   Florida umbrellas
20   Bermuda high back patio chairs
20   Florida high back sprung chairs
15   Florida auto leg rest chairs
15   Florida sprung loungers

1 Work out the total cost of Tony's order.
   Try not to use a calculator.

   Reminder
   $10 \times £11.99$
   $= 10 \times £12 - 10 \times 1p$
   $= £120 - 10p$
   $= £119.90$

2 The manager of the supermarket gave Tony a
   10% trade discount.
   Work out the amount Tony actually paid for the
   furniture.

3 This is a plan of the swimming pool area in the
   leisure centre. ▶
   Some of the furniture Tony bought is to be
   arranged around the pool.
   Make a good scale drawing to show how much
   of the furniture could be placed here without
   overcrowding.

4 After a few months in business Tony decides to
   increase the overall patio area (which includes
   the swimming pool) to 96 m².
   The swimming pool will remain the same size.

   Draw a scale plan of the new patio area.
   Show how you would arrange the furniture
   (your new plan should have more furniture).

5 Tony also wants to tile the new patio area with
   tiles measuring 20 cm by 20 cm.
   (a) Work out how many tiles he will need.
   (b) Tiles are sold in packs of 50.
       Each pack costs £37.50.
       How many packs of tiles are needed, and
       how much will they cost?

Florida Umbrella
£12.99

Florida High Back Sprung Chair
£4.99

£11.99 Bermuda Round Patio Table 93cm

£10.99 Bermuda High Back Patio Chair

Florida Sprung Lounger
£11.99

Florida Auto Leg Rest Chair
£19.95

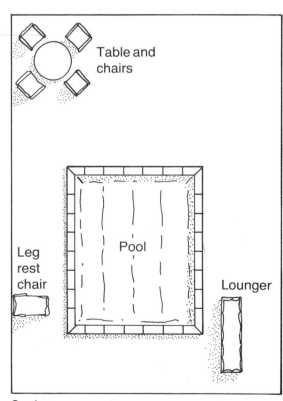

Table and chairs

Leg rest chair

Pool

Lounger

Scale:
1 cm represents 1 m

Temperature graphs

These tables are taken from the weather page of a newspaper.

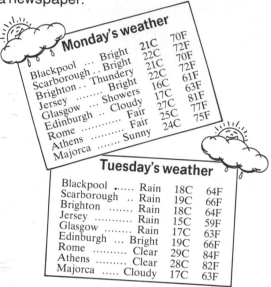

Monday's weather

Blackpool ...	Bright	21C	70F
Scarborough ..	Bright	22C	72F
Brighton ..	Thundery	21C	70F
Jersey	Bright	22C	72F
Glasgow ...	Showers	16C	61F
Edinburgh ..	Cloudy	17C	81F
Rome	Fair	27C	77F
Athens	Fair	25C	75F
Majorca	Sunny	24C	

Tuesday's weather

Blackpool	Rain	18C	64F
Scarborough ..	Rain	19C	66F
Brighton	Rain	18C	64F
Jersey	Rain	15C	59F
Glasgow	Rain	17C	63F
Edinburgh ...	Bright	19C	66F
Rome	Clear	29C	84F
Athens	Clear	28C	82F
Majorca	Cloudy	17C	63F

1 Draw a graph to show both Monday and Tuesday's temperature (in °C) at each place mentioned in the tables.
Step 1
Get a sheet of A4 size, squared paper (½ cm squares).
Step 2
Turn the sheet sideways.
Step 3
Mark out the scales as shown below.
Your graph should go up to 30 on the temperature scale.
Step 4
Draw 2 columns for each place in the lists.

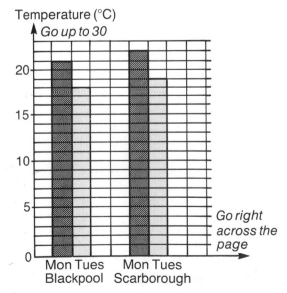

Temperature (°C)

Go up to 30

Go right across the page

Mon Tues Mon Tues
Blackpool Scarborough

2 From your graph make 3 lists.
- (a) Places which had a *fall* in temperature from Monday to Tuesday.
- (b) Places which had the *same* temperature both days.
- (c) Places which had a *rise* in temperature from Monday to Tuesday.

3 Draw a graph which converts °C into °F.
Step 1
Copy and complete this table.
Use the information in the two weather tables.

°C	15	16	17	18	19	21
°F	59				66	
°C	22	24	25	27	28	29
°F						

Step 2
Use the other side of your A4 sheet of paper, turned this way round.
Step 3
Mark out the scales.
Mark °F up the side.
Go from 0 to 100 in steps of 10.
Mark °C along the bottom.
Go from 0 to 40 in steps of 5.
Step 4
Plot each pair of numbers from your table on to the graph. The first point is at (15°C, 59°F).
T Ask your teacher for help with this part.
Step 5
The points are almost all in a straight line.
Draw the best straight line you can through the points.
Make it go right across the page.

4 Now answer these questions using your graph.
- (a) Change 10°C into °F.
- (b) Change 35°C into °F.
- (c) Change 40°F into °C.
- (d) Change 60°F into °C.
- (e) Change 100°F into °C.
- (f) Water freezes at 0°C. What temperature is this in °F?

5 This table shows the temperature in Glasgow each day for two weeks in December.

Temperature (°C)

Sun	Mon	Tue	Wed	Thu	Fri	Sat
10	5	0	−3	−6	−2	0
Sun	Mon	Tue	Wed	Thu	Fri	Sat
0	3	−5	−8	7	3	−5

(a) Copy and complete the line graph below to show these temperatures. Your graph should go right across the page. It should have all 14 days on it.

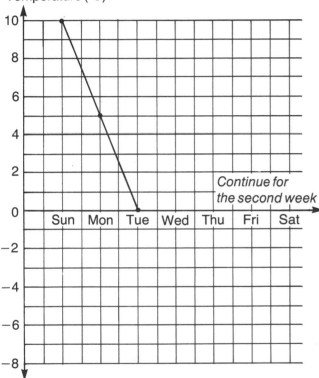

(b) Copy and complete the table below. Your table will have 13 lines in it.

Temperature difference (Glasgow) (°C)

Sun – Mon	Fall of 5°
Mon – Tue	Fall of 5°
Tue – Wed	
Thurs – Fri	
Fri – Sat	

(c) Between which 2 days was there the biggest change in temperature?

6 This is the temperature difference table for Liverpool for the same two weeks.

Temperature difference (Liverpool) (°C)

Sun – Mon	Fall of 2°
Mon – Tue	Fall of 3°
Tue – Wed	Fall of 3°
Wed – Thu	Fall of 1°
Thu – Fri	Rise of 1°
Fri – Sat	Rise of 4°
Sat – Sun	Fall of 1°
Sun – Mon	No change
Mon – Tue	No change
Tue – Wed	Fall of 3°
Wed – Thu	Fall of 1°
Thu – Fri	Fall of 3°
Fri – Sat	Rise of 4°

(a) Make a table to show the temperature in Liverpool on each of the 14 days. (The temperature on the first Sunday was 5°C.)

(b) Draw a line graph to show the Liverpool temperatures.

(c) 'Glasgow was colder than Liverpool during these 2 weeks.' Write down your comments on this.

7 This table shows the maximum and minimum temperatures in London during a cold spell.

	Min temp	Max temp	Range
Mon	−1°	4°	5°
Tues	−4°	0°	
Wed	−3°	5°	
Thurs	0°		2°
Fri		−1°	8°
Sat	−5°		7°
Sun	−2°	2°	

Copy and complete the table.

Chinese carpets

CHINESE CARPETS
An outstanding special purchase

A beautiful collection of hand made carved carpets, extra fine quality in a range of superb colours and designs at exceptional prices.

10ft × 8ft
When perfect £1,399   SECONDS **£699**

12ft × 9ft
When perfect £1,799   SECONDS **£899**

14ft × 10ft
When perfect £2,399   SECONDS **£1199**

This advert describes the 3 sizes of Chinese carpet available. You will have to work out the area for each size.

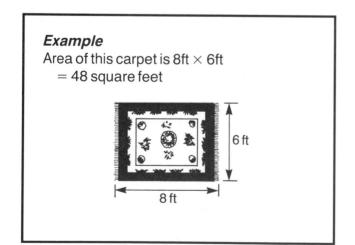

Example
Area of this carpet is 8ft × 6ft
= 48 square feet

6 ft

8 ft

1 Work out the area of the first carpet in the advert.

2 How much does each square foot of this carpet cost?

3 Repeat questions 1 and 2 for the other two carpets in the advert.

4 What does 'seconds' mean?

5 Work out the *perfect* price per square foot of each carpet.

6 Copy the table *below* and complete it to show the answers you have worked out.

Size of carpet	Perfect price per square foot	Seconds price per square foot

7 These Chinese carpets are very expensive, so you do not fold them or cut them up. What is the biggest size of carpet that would fit into each of these rooms? (Rooms (c) and (d) will need two carpets.)

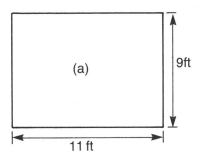

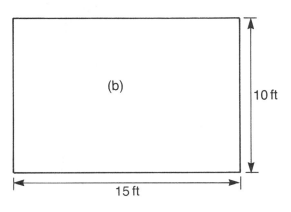

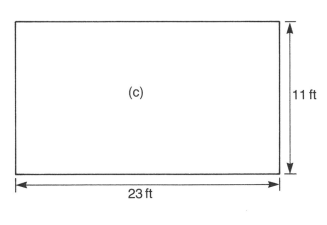

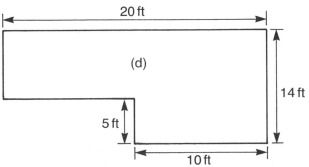

8 Draw a neat sketch to show how the carpet(s) fit into each of the rooms in question 7.

9 This is a scale drawing of a large sitting room.

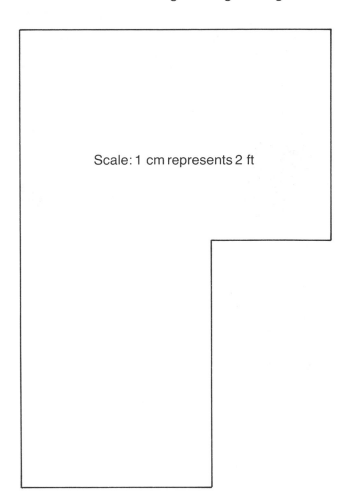

Scale: 1 cm represents 2 ft

(a) Draw each of the carpets in the advert on card to the same scale as the plan above. Cut out each shape.

(b) Use your cut-outs to find 2 ways of covering the sitting room floor with the Chinese carpets. Which way costs less? By how much?

10 Use your cut-outs to solve this problem. How could a rectangular room measuring 22 ft by 20 ft be covered with Chinese carpets, leaving as little space as possible between carpets?

Nurses' pay

In this section you should round amounts of money to the nearest £100.

> ### Examples
> £2798 rounded to the nearest £100 is £2800
> £3325 rounded to the nearest £100 is £3300
> £7186 rounded to the nearest £100 is £7200
> £9149 rounded to the nearest £100 is £9100

1 Round each of these amounts to the nearest £100.

(a) £3789   (b) £4579   (c) £1670
(d) £2525   (e) £3276   (f) £9535
(g) £6056   (h) £6048   (i) £3915
(j) £3987   (k) £12 390   (l) £10 964

2 This newspaper article is about nurses' pay. Read the article carefully.

BOTTOM OF THE PAY HEAP

Compared with other public service professions, nurses are at the bottom of the pay heap.

Student/pupil nurses have a gross starting salary of **£3695**. A staff nurse gets **£4998** and a sister **£6321**. It takes a student three years to get to staff nurse status and five years to reach sister.

Teachers in primary schools start on **£6006**, rising to **£9378** after 13 years' service. Their secondary school colleagues begin on **£6150** and reach between **£9435** and **£9885** after 12 years.

Police can earn from **£6708** to **£11,193** as a constable after 15 years' service. Sergeants get between **£10,704** and **£12,282**, inspectors are on **£12,282** to **£13,944** and chief inspectors are paid between **£13,944** and **£15,513**.

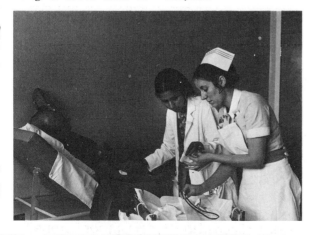

3 The article in question 2 contains a large amount of information.
It would probably be clearer on a graph.

(a) Copy and fill in each of these tables. Round each salary to the nearest £100. When the article gives two salaries you should choose the lowest.

Nurse

Years	0	3	5
Salary (£)			

Primary teacher

Years	0	13
Salary (£)		

Secondary teacher

Years	0	12
Salary (£)		

Police constable

Years	0	15
Salary (£)		

(b) You are going to draw a graph to show these salaries (all on the same grid). Draw the axes shown below on ½ cm squared paper. Your graph should go up to £12 000 for Salary and 15 for Time in years.

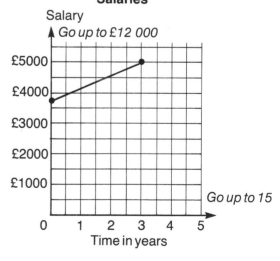

Salaries

(c) Mark a set of points on the graph for each of the four salaries.

(d) Join up each set of points with a straight line. (If you have coloured pencils, make each line a different colour.)

4 Study your graphs carefully.
Write down what you think the point of the article was.

Sports news

The Ryder Cup is a golf competition played between a European team and an American team.

The results of the 1985 Ryder Cup matches are shown below.

European players are mentioned first in each match.

Scoring
Win 1 point   Lose 0 points   Draw $\frac{1}{2}$ point

THE RYDER CUP 1985

First day foursomes

S. Ballesteros and M. Pinero beat
   C. Strange and M. O'Meara ....
N. Faldo and B. Langer lost to
   C. Peete and T. Kite ....
S. Lyle and K. Brown lost to
   L. Wadkins and R. Floyd ....
H. Clark and S. Torrance lost to
   C. Stadler and H. Sutton ....

First day fourballs

P. Way and I. Woosnam beat
   F. Zoeller and H. Green ....
Ballesteros and Pinero beat
   P. Jacobsen and A. North ....
Langer and J. M. Canizares and
   Stadler and Sutton drew ....
Clark and Torrance lost to
   Wadkins and Floyd ....

Result – Europe ■ , United States ■

Second day fourballs

Way and Woosnam beat
   Green and Zoeller ....
Torrance and Clark beat
   Kite and North ....
Ballesteros and Pinero lost to
   O'Meara and Wadkins ....
Langer and Lyle and Stadler and
   Strange drew ....

Second day foursomes

J. Rivero and Canizares beat
   Kite and Peete ....
Ballesteros and Pinero beat
   Stadler and Sutton ....
Way and Woosnam lost to
   Strange and Jacobsen ....
Langer and Brown beat
   Floyd and Wadkins ....

Result – Europe ■ , United States ■

Third day singles

Pinero beat Wadkins ....
Woosnam lost to Stadler ....
Way beat Floyd ....
Ballesteros and Kite drew ....
Lyle beat Jacobsen ....
Langer beat Sutton ....
Torrance beat North ....
Clark beat O'Meara ....
Faldo lost to Green ....
Rivero lost to Peete ....
Canizares beat Zoeller ....
Brown lost to Strange ....

Result – Europe ■ , United States ■

Overall result – Europe ■ , United States ■

---

1 Use the scoring system above the table to work out the points scored by Europe and America on:
  (a) the first day
  (b) the second day
  (c) the third day

2 Use your answers to question 1 to work out the overall result of the competition.

3 How many points are needed to win the Ryder Cup? (Count the total number of matches.)

4 Work out the total score at the end of the second day.

5 The third day's matches were played in the order shown in the results table.
Which European player scored the points which clinched the Ryder Cup for Europe?

Small ads

Linda put this ad in the local paper. ▼

POLAROID camera with flash attachment and case. £22. SLR kit comprising Pentax Super, 3 lenses, case and tripod. £150. Northbank 7561.

The cost of the ad was £9.50. Linda had to fill in the form opposite. ▶

SMALL ADS

Polaroid	camera	with	flash
attachment	and	case	£22.
SLR	kit	comprising	Pentax
Super,	3	lenses	case

Minimum £7.00

and	tripod	£150.	Northbank
7561			

50p for each additional word

Fill in a blank form for each of the adverts below.
Write down the cost of each one.

1
CAR ALARMS supplied and fitted from £40 including siren. Fully guaranteed. Briggs Auto Electrics Limited. 231 742 1235. Open seven days.

(The phone number counts as 3 words.)

2
HAIRSTYLIST for gents salon. Must be able to do traditional and modern styling. 556 7840.

3
FREE full time or part time courses available in shorthand and typing. For further details telephone or call Touchtype, 137 Bath Street, Eastfield 7056.

4
AVENGER 1979, V registration 34 000 miles, excellent bodily and mechanically, tow bar, radio, stereo cassette, £1250 ono. 751 8677.

5
RE-COVER all types of suites. Fantastic range of dralons, velours, tweeds, etc. Estimates free. Fully guaranteed. Big discount for cash. Rep will call day or evening, all districts. Phone 085 7554 day or 085 3196 evening.

6
FORD Cortina 1600cc 1975, four months' MoT. Six months' tax, excellent car for year, great driver, 28mpg. Extras include radio, alarm, sports wheels, front spots, rear heater, leopard skin covers, £450 ono. 123 587 199.

Make up adverts for each of these. Say how much the cost is.

7 You want to sell a 1974 Ford Escort 1300. It has done 55 000 miles and you bought it new. You want around £300. It has had a new exhaust fitted recently and 4 new tyres.

8 You want a job. Say something about yourself, especially your good qualities. Say what kind of job you would like and why an employer should give you one.

Your teacher will help you with this section.

9 You are going to draw a graph to show what each size of advert costs.
You need a large sheet of $\frac{1}{2}$ cm squared paper.
Part of the graph is shown here to help you.

Cost of small ads (£)

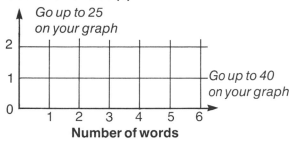

Go up to 25 on your graph

Go up to 40 on your graph

Number of words

(a) Mark a point on the graph for each number of words. (If you see the pattern you do not have to mark them all.)
(b) Join up all the points using a ruler.
Show your graph to your teacher before you go on.

10 This table shows the usual number of adverts in the paper each week.

Number of words	Cost	Number of ads	Income
16 or less	£7	15	£105
17		8	
18		12	
19	£8.50	8	£68
20		4	
21		2	
22		3	
23		2	
24		1	
25		5	
26		3	
27		6	
28		0	
29		0	
30		3	
31		4	
32		1	
33		2	
34		4	
35		0	
36		1	

Copy the table and fill it in. Use your graph to find the costs.
Work out the total income from the small ad pages each week.

The sales manager of the paper decides to change the order form like this:

SMALL ADS

Minimum £10

25p for each additional word

11 Draw a graph to show the new prices on the same grid as the graph for question 9.
Make it a different colour.

12 Answer these questions from your graphs.
(a) What size of advert will cost less with the new prices?
(b) What size of advert will cost more with the new prices?
(c) What size of advert costs exactly the same?

13 The manager expects to get the same number of adverts as before. (See question 10.)
(a) Work out the expected income from the small ad pages at the new prices.
(b) What could happen to disappoint the sales manager?

14 Here are the accounts of the small ad page for the first 8 weeks of the year.

Week	Income (£)
1	850
2	825
3	800
4	785
5	800
6	750
7	710
8	740

(a) Draw a line graph to show these figures.
(b) If you were the sales manager, how would you feel about this?
What would you do about it?